DDT Wars

DDT Wars

Rescuing Our National Bird,

Preventing Cancer, and Creating

the Environmental Defense Fund

CHARLES F. WURSTER

OXFORD
UNIVERSITY PRESS

OXFORD
UNIVERSITY PRESS

Oxford University Press is a department of the University of Oxford.
It furthers the University's objective of excellence in research, scholarship,
and education by publishing worldwide.

Oxford New York
Auckland Cape Town Dar es Salaam Hong Kong Karachi
Kuala Lumpur Madrid Melbourne Mexico City Nairobi
New Delhi Shanghai Taipei Toronto

With offices in
Argentina Austria Brazil Chile Czech Republic France Greece
Guatemala Hungary Italy Japan Poland Portugal Singapore
South Korea Switzerland Thailand Turkey Ukraine Vietnam

Oxford is a registered trade mark of Oxford University Press
in the UK and certain other countries.

Published in the United States of America by
Oxford University Press
198 Madison Avenue, New York, NY 10016

Library of Congress Cataloging-in-Publication Data
Wurster, Charles F.
  DDT wars : rescuing our national bird, preventing cancer, and creating the Environmental
Defense Fund / Charles F. Wurster.
     p.   cm.
  Includes bibliographical references and index.
  ISBN 978-0-19-021941-3 (alk. paper)
  1. DDT (Insecticide)—Toxicology. 2. DDT (Insecticide)—Physiological effect. 3. DDT
(Insecticide)—Environmental aspects. 4. Environmental Defense Fund. 5. Health promotion.
6. Environmental health. I. Title.
  SB952.D2W87 2015
  632'.9517—dc23
                          2014042820

9 8 7 6 5 4 3 2 1

Printed in the United States of America on acid-free paper

Dedicated to Lorrie Otto,
who did her best to protect life on Earth,

to William D. Ruckelshaus,
a public servant who did the right thing,

to Marion Lane Rogers
who kept this history alive and inspired us all,

and to Marie H. Gladwish,
who more than anyone else, with endless encouragement,
affection and her many talents, helped bring this book to fruition.

CONTENTS

The Bald Eagle, our national bird, engraved on the Great Seal of the United States of America, was disappearing from America. Predatory birds of many species were in sharp decline. Brown Pelicans in California laid eggs that broke and produced almost no chicks. Mother's milk was so contaminated that if it were in any other container, it could not legally cross state lines. Humans worldwide were contaminated, as were penguins in Antarctica and polar bears in the Arctic. Laboratory tests indicated that the insecticide DDT caused cancer. William Ruckelshaus, first administrator of the U.S. Environmental Protection Agency (EPA), banned DDT in 1972.

Told by someone who lived it, this book details how a modest group of volunteer scientists and citizens fought the "DDT wars" from Long Island living rooms through the courts to ultimate victory. This was overwhelmingly a grassroots effort. They were not revolutionaries, did not demonstrate in the streets, threatened nobody, and hugged no trees; nobody lay down in front of a bulldozer. They had neither wealth nor political connections. What they did have were determination, passion, and persistence to eliminate the DDT threat. They got the science right.

This dedicated group of scientists, citizens, and a few attorneys escalated the DDT issue from local to state to national prominence. They used our democratic, legal, and political systems designed for the peaceful resolution of disputes. Experts from varying disciplines from around the country and the world rallied to the cause. Margaret Mead said, "Never doubt that a small group of thoughtful, committed citizens can change the world. Indeed, it is the only thing that has." That is exactly what happened here.

Along the way a new organization, the Environmental Defense Fund (EDF), was created, without funds, by 10 citizens, the same volunteers who pursued the DDT campaign. Their strategy was to take environmental problems to court using scientific evidence. Significant legal obstacles challenged the campaign from the beginning. The team was repeatedly thrown out of court.

Rachel Carson's groundbreaking book *Silent Spring* aroused public opinion, caused substantial alarm, and was vigorously denounced by the chemical industry. Her analysis of the scientific literature was accurate, but *Silent Spring* did not immediately change government pesticide policies. EDF, through a systematic and persistent campaign employing scientific evidence through legal channels, did change government pesticide policies.

By chance and circumstances, I found myself in the midst of this course of events beginning in 1963. I was told I should write a book about it, but why write a book 40 years after the fact? Why now? When it seemed timely and sensible 40 years ago, when the whole story was fresh in my mind and all the relevant papers surrounded me, I retreated from the task. Many other things needed to be done, and the job was intimidating. Somebody else would write this book. Furthermore, the goal had been to get DDT banned, and that had been accomplished. The fledgling organization EDF still needed nurturing and attention; so did my full-time, paying faculty position at the State University of New York at Stony Brook. The DDT battles paid nothing. If anything, they were a detriment to my career. Writing a book just did not fit in at that time.

During the past decade my thinking began to change. "Somebody else" had not written this book. An accurate account of the DDT saga did not exist. Distortions began to appear. The Internet contained all sorts of false nonsense about why and how DDT got banned. A TV documentary on the Bald Eagle stated that "Congress" had banned DDT. Internet websites asserted there never was a decline of the birds. The DDT ban was based on "junk science" and the DDT threat was a "hoax." I was called "a radical ecologist out to destroy the American way of life." Scientists were actually called "genocidalists," killing millions with malaria. DDT had become "controversial" again. The same tactics that were used back in the Sixties to denounce the DDT threat were again being used to discredit scientists and the scientific evidence of climate change. The public was being misled and deceived again.

Maybe it was time to write the book.

Like falling dominoes, enduring benefits from winning the "DDT wars" continued to roll in. The bird populations that had been depressed by DDT had recovered their former abundance. The Bald Eagle had returned dramatically. First a dozen, then two dozen dangerous toxic chemicals were banned worldwide. Environmental law spawned by the DDT precedent had become a significant part of the legal profession. EDF had become one of the most influential and innovative environmental organizations in the United States and had even gone international.

The original intention to save birds from the ravages of DDT soon included the additional goal of preventing human cancer. The DDT campaign by a growing group of scientists and citizens had made the world a safer, healthier place. That is EDF's continuing focus.

What an exciting and dramatic sequence of events! Sometimes discouraging, sometimes exhilarating, always demanding intense involvement, we started with modest goals but attained significant accomplishments. This journey, these results, is among the greatest environmental case histories of modern times. It heralded a new era of environmental defense. This 50-year story is a book that could not have been written 40 years ago.

Now is the time to write this book!

This book details the inside story of the DDT wars as the struggle escalated, roughly in chronological order, from about 1963 into the 1970s. New issues were spawned, and some of these have been followed to more recent times. This is not a peer-reviewed scientific treatise, but every effort has been made to achieve scientific accuracy while still being comprehensible to intelligent laymen. Where possible, reviews are cited that could lead readers to original sources, rather than citing hundreds of individual papers. Scientific information available at the time of these events is cited. We will not "bring everything up to date," but the science of the DDT issue is even truer now than it was then. Read on to see how the DDT wars of 40 years ago became important in today's world.

Fifty years after Rachel Carson published her book *Silent Spring*, and 40 years after the U.S. Environmental Protection Agency banned DDT, the pesticide is still controversial, although it shouldn't be.

Dr. Wurster's book is a wonderful, thought-provoking narrative of how dying birds inspired the fight to ban DDT, helping to spawn the American environmental movement. This movement took on many other battles, including cleaning up the nation's waterways, providing safe drinking water and food, and cleaning the air. Today, it is dealing with dangerous greenhouse gases that threaten the health of the entire planet.

I served at the EPA as assistant administrator in the 1990s and was responsible for EPA's pesticide programs. During my time at EPA it was clear that the successful efforts by a small number of volunteers to ban DDT had continued to bear fruit. Not only had most uses of DDT been eliminated, but a number of other persistent toxic pesticides had also been banned. The laws governing pesticide approvals in the United States were greatly strengthened overall as a clear consequence of the important scientific and legal precedents that were established. We now have treaties—the Stockholm Convention, a global agreement to control the release of persistent organic pollutants (the POPs Treaty), and the Rotterdam Convention on Prior Informed Consent—that are helping to control the most persistent and hazardous toxic pollutants everywhere. The benefits from these efforts to health and the environment are enormous.

A 2011 review by the World Health Organization (WHO) found that people in households treated with DDT for the prevention of malaria, and the workers who apply DDT, are at risk for cancer and male reproductive defects. Even public health uses of DDT need to include safeguards for workers and households. In homes, women of childbearing age, fetuses, and infants are especially susceptible. In 1972, EPA Administrator Ruckelshaus decided there was enough scientific evidence to warrant banning most uses of DDT in the United States. DDT toxicity was clearly demonstrated in the 1970s, and the evidence today, after 40 more years of research, is even stronger. In truth, we know more about DDT than almost any other chemical.

Why would DDT be controversial today? For many, DDT is a symbol of the environmental movement. From the very beginning, while a small number of volunteers were waging battles to decrease the risks of DDT, there were many defenders of DDT, some paid by industry to refute the scientific evidence. Even today, some conservative voices continue to attack the DDT ban, often claiming that it caused deaths due to malaria. Yet to my knowledge, no national or international legal actions ever have taken DDT out of the hands of public health authorities for controlling the mosquitoes that carry malaria. If anything, malaria control programs have suffered from political instability, health system weaknesses, and funding shortages, not lack of available pesticides. In many regions of the world mosquitoes have evolved immunity to DDT, rendering it ineffective. In Mexico, malaria has been controlled in this century without the use of any DDT.

So why is there still an attack on the issue of DDT today? As in the 1970s, we today are locked in a debate about a major environmental issue, climate change, that threatens our planet and life as we know it. Many people are understandably confused about the science of climate change. In the past, industry attacked the scientific evidence on the risks of DDT. Today, consultants and academics paid by industry-supported funding are attacking the science of climate change. They may hope that by claiming that the DDT threat was a "hoax," the DDT ban killed millions, and the science was inconclusive, they can strengthen their case against the science connecting manmade emissions of greenhouse gases to climate change. Or, at the very least, they can succeed in sowing doubt and confusion.

There is much to learn from understanding the past. These hard-fought battles over pesticide safety were not won overnight. The issues were every bit as complex (at the time) as our current efforts to confront global climate change and, like today, required that scientists, attorneys, and policy experts come together to develop effective strategies. While in the past a small number of committed volunteers were able to make an enormous difference, today we need many more allies to prevail. The efforts to discredit the science on DDT risks connect directly to the efforts to discredit the science of global warming.

The actions on DDT spurred by Dr. Wurster and others, whose initial intention was to save birds, ultimately also reduced human exposures, which probably prevented cancer, along with human developmental and reproductive defects. Those benefits to birds and to human health are still with us today and will carry forward into the future. Let us hope we are able to continue to make progress in environmental health as we confront the challenges of today's world.

Lynn R. Goldman, MD, MPH, MS
Michael and Lori Milken Dean of Public Health
Milken Institute School of Public Health
The George Washington University
Washington, DC

ACKNOWLEDGMENTS

The DDT story of 40 years ago is about science, law, and people, a remarkable and growing group of dedicated people who wanted to protect our environment. Most were not paid to pursue this mission but were eager to be a part of it. That includes all those scientists and other citizens mentioned in this book, and many more unmentioned. It especially includes a growing list of scientists called EDF's "Scientists Advisory Committee" who agreed in writing above their signatures to provide information or testimony in their areas of expertise without fee. I often called complete strangers to ask them to come testify on short notice, and I was amazed that they usually dropped whatever else they were doing to fly across the continent to testify in a legal proceeding.

Then there was the EDF team, incredibly dedicated and talented people, starting with only 10 and growing into hundreds, then thousands, who weathered the ups and downs of building an organization, starting with nothing. We often didn't know how to do it, made all the mistakes possible, and usually gained some wonderful new person who did know how to do it. There were often good reasons to quit, but few did, and many dedicated decades of their lives to building the EDF as we know it today. All of this was a team operation. I was urged to write this book in the first person, but I will usually say "we" because it was almost always "we," a team effort. Special thanks are due to Bob Risebrough and Ian Nisbet, who took many weeks from their usual activities to lend their great scientific expertise to the hearings in Madison, Wisconsin, and Washington,

DC. Tom Cade of The Peregrine Fund helped make sure I got the details right in my treatment of predatory birds.

It will be difficult to mention individuals because I will invariably overlook many deserving of mention. Art Cooley and Malcolm Bowman kept after me for years to write this book. Marion Lane Rogers, the EDF secretary, hassled me to contribute to her *Acorn Days*, personal recollections of those early days, before I finally did so while in Australia. Marion vigorously urged me for years to write this book, and encouraged me to quote from *Acorn Days* without limit. Vicki Eisel, Debbie Mefferd, and Norma Watson, as editors of the EDF newsletter, gathered and organized detailed information about EDF's early days. Norma supplied me with two large notebooks containing all of EDF's newsletters and other publications since day one. *Acorn Days* and those EDF newsletters were major sources for this book. Norma also helped edit the manuscript. My former wife, Eva Tank-Nielsen, put up with the emotional gyrations of the 1970s while encouraging me not to quit when there were good reasons to quit.

My partner, Marie Gladwish, joined the others in urging me to write it, offered no end of encouragement, and provided many excellent ideas, helpful editing, and graphic designs and photography. My son, Erik Wurster, repeatedly got me out of computer trouble. When my format got scrambled, I sent the manuscript to him and he sent it back unscrambled. Most of the time he was in Rwanda, East Africa, but I still sent it to him and it was quickly back again, often within a half-hour. Once I touched the wrong key and everything vanished in an instant, gone directly to heaven. Petrified, I called Erik, and he brought it back from heaven in minutes. Malcolm Bowman also frequently bailed me out of digital headaches, patiently walking me through computer procedures on late-night, transcontinental telephone calls.

I thank Laura Chittenden, who served as an editor of the early manuscript, and David Bjerklie, who was the fact checker of the later edition. A lengthy dialogue with Art Cooley, Bev Grant, and Marie Gladwish sharpened my focus in weaving this story together. I also thank The Peregrine Fund for supplying electronic copies of numerous references, which avoided many trips to the library. Kristen Nyitray and Lynn Toscano were most helpful as managers of the invaluable EDF Archives in the Special Collections at Stony Brook University. Joel Plagenz and Cynthia Hampton of EDF's staff supplied logistical and moral support when the hurdles seemed high. Bill Berry helped me wend my way through the thicket of securing photos and finding copyright holders, an endeavor I had not encountered before.

Special thanks go to Rod Cameron and Bill Butler for providing detailed impressions in their own words of those early days when EDF might have disappeared, and to Jim Moorman and John Dienelt for insisting that my descriptions of EDF's federal litigation were accurate. I particularly want to thank Ian Nisbet for providing scientific support and writing the sections on aldrin, dieldrin, heptachlor, and chlordane in Chapter 12. I sincerely thank Dr. Lynn Goldman and EDF President Fred Krupp for their important contributions to this book. Lastly, I appreciated the patience and prompt replies to my questions by Jeremy Lewis and others at Oxford University Press, from an author who never wrote a book before and will not again.

DDT Wars

A New England Town Sprays Its Elm Trees with DDT

The robin was twitching, tremoring, convulsing uncontrollably, and peeping occasionally. The student handed the bird to me, and in a few minutes it was dead in my hands. It was April 23, 1963, and I was in my laboratory at Dartmouth College in Hanover, New Hampshire, when the student walked in with the bird. A week earlier the elm trees of Hanover had been sprayed with the insecticide DDT to control the spread of Dutch elm disease by elm bark beetles.

In the following weeks 151 dead birds filled my freezer, many of them exhibiting before they died the tremors that we later learned were typical of DDT poisoning. Four of us were conducting a small-scale study of the effects, if any, of the DDT spray program in Hanover. We were shocked by what was happening to the local birds, but we would have expected this reaction to DDT if we had read the scientific literature on earlier DDT spray programs on elm trees. We had not.

We soon realized that we had rediscovered what other ornithologists had already reported from DDT spray programs in the American Midwest. We also soon learned that DDT was ineffective in preventing the spread of Dutch elm disease and that another procedure, sanitation without insecticides, effectively protected the elms. This DDT spray procedure was all costs and no benefits. Hundreds of towns were killing thousands or millions of birds while not protecting their elms. The whole thing struck me as absurd and tragic. It became a life-changing event for me. I decided that DDT was a chemical that had to be stopped, although I hadn't the slightest idea where such a conclusion was going to lead.

HOW DID I GET INTO THIS?

I was 33 years old and had become what in those days was usually called a conservationist. Now such people have been renamed "environmentalists." I had a dubious beginning as such a person. When I was about seven and living in a northern suburb of Philadelphia, I came across a couple of snakes. I beat them to death with rocks! I guess I had already "learned" that snakes are "bad." I soon developed an interest in snakes at a summer camp in the mountains of Pennsylvania as a teenager. It was there that I caught and skinned a number of venomous copperheads. Those snakes made out no better than those original two garter snakes, which they must have been. I also collected box turtles and painted turtles, which I kept as pets for years. They made out much better than had the snakes. I released the turtles back into the wild.

By the time I was 10 I had learned to play the trumpet, so I became the camp bugler, blowing 22 bugle calls per day to keep the camp on schedule. That gave me two months each summer at a boy's camp for free in the Pennsylvania mountains, which contributed much to my knowledge of natural history. I also got interested in guns (and firecrackers), and over several years I earned the rank of "Expert Rifleman" from the National Rifle Association. I still have the medals to prove it. When back in Philadelphia, I sometimes wandered around the neighborhood shooting

House Sparrows and European Starlings with my BB gun. I told myself they were pest birds, so maybe I was doing some kind of public service. Looking back on it, I guess I was somewhat out of control. In the late 1940s when I was about 17 and in high school at Germantown Friends School in Philadelphia, I met a biology teacher, naturalist, and outstanding ornithologist named Joe Cadbury, who had a huge influence on me and permanently changed my life. He had a tradition of taking a car full of kids to southern Florida for 10 days during spring vacations. I went on several of these excursions. Most of the group was birding while I slopped around in the Everglades to catch a few snakes. I brought the snakes back to Philadelphia in cloth bags and gave them to the Philadelphia Zoo in return for free entrance tickets. At one point I had several large venomous water moccasins in barrels in my cellar at home. I don't recall what I was going to do with them, but they generally had nasty dispositions and were not especially pet-like. One morning I was shocked to find all of them dead. About 25 years later my father told me he had dropped rat poison into the barrels. Snakes were apparently OK with my parents, but not large and aggressive venomous ones.

One snaky episode remains in my memory. We were sleeping on the beach on Sanibel Island, then undeveloped, in southwest Florida. Biting sand flies were driving us crazy and sleep was impossible, so all six of us got into the big old Packard and the other five smoked cigarettes (excluding me—I never smoked) to fill the car with smoke and choke the sand flies. I guess it worked, but suddenly the car seemed filled with snakes and we all jumped out in a panic.

I had caught various snakes, including at least one venomous water moccasin, and the cloth snake bags were hanging by their knots from the strap on back of the front seat. A cigarette spark had fallen onto the bag and made a hole large enough for a snake to escape. Nobody had noticed the smoldering bag, since the car was already full of smoke. It was pitch dark, so nobody knew which snake or snakes had gotten out of the bags. It turned out that it was a six-foot-long but harmless chicken snake. I put the chicken snake into a new bag, but my snake-catching activities became less popular with my friends.

On the Florida trips I soon realized that in a day the others would see more than 100 species of birds while I caught a few snakes. I seemed to be on the short end of the stick. Birds were vastly more numerous, diverse, and interesting. Birds were where the action was. Without any effort, by osmosis through much time with those birders, I had already learned about 200 bird species. That did it. I became a birder for life. In the following 60 years I have seen about 4,000 avian species on all continents, and the DDT issue brought birds into my professional life as well. Maybe I can claim the title "ornithologist."

It was in Florida that I became familiar with the Bald Eagle, our national symbol, a majestic species if ever there was one. Adults are unmistakable with their white head and tail. It was 1947 and they were fairly common in Florida then. I did not know at the time that they would suffer a large decline in numbers in the next two decades, that they would become very scarce nationwide, and that it would be 1970 before scientists had figured out the cause of the decline. The story of that iconic keystone species at the top of its food web will develop as we move along in our chronology of events.

By 1950 I had also become familiar with other bird species that I had no idea were beginning to decline. Ospreys were common in Florida and along the East Coast, their large nests on poles and bare trees for anyone to see. We often saw them carrying fish to feed their young. Incredibly, when they catch a fish they reorient it in their talons so that it is carried head first, thus reducing air resistance as they fly back to their nest.

Brown Pelicans were abundant in Florida, often playing follow-the-leader in lines over the surf, clownishly entertaining people along the beach. Peregrine Falcons were not common but were occasionally seen in a wild chase through the sky after a panicked shorebird. It would be another decade before anyone knew that something was the matter with these and many other species of birds.

I retain a passing interest, indeed somewhat of an appreciation, for snakes. During one lunch hour in the hills above Stanford University, where I was attending graduate school about 1956, I came across a two-foot Pacific rattlesnake. I brought him back to my apartment in my brown paper lunch bag, and he lived in a large bell jar for a year, consuming one white mouse per week. Named "Pretty Boy," he was gentle, easily handled, and ceased to rattle because he was no longer frightened. He was allowed short outings on the living room floor. This performance spooked an insurance salesman one afternoon, who quickly departed without selling any insurance.

Proving that I was a true conservationist by this time, I then took Pretty Boy back to where I had found him, and he was pleased to be free again. I had developed considerable admiration for rattlesnakes by appreciating the extremely complex and sophisticated chemistry of their venom systems, and the brilliant stereochemoreception of their forked tongues for following and consuming rodents. When they inject venom into their prey they also take an imprint of its odors and chemistry, and their forked tongue tells them whether to turn right or left in following it to its demise. It's too bad that people persecute them, killing rattlesnakes on sight or rounding them up by the hundreds and thousands to be slaughtered in weekend social events. Admittedly, rattlesnakes are not good backyard playmates for small children, but decreasing the rattlesnake population merely leads to more rats.

After high school and my somewhat dubious beginning as an environmentalist, my pathway became a bit more conventional. I attended Haverford College and the University of Delaware, I married Doris Hadley, and we both went to Stanford University to complete doctorates, hers in biology and mine in organic chemistry, in 1957. I had a Fulbright Fellowship in Innsbruck, Austria, for a year, and we then spent eight months wandering by land across the Middle East to India, Kashmir, Nepal, and Southeast Asia, then to Japan and ultimately San Francisco.

In 1959 I took a job with Monsanto Research Corporation north of Boston, where my research involved jet fuels and laminating resins. But my interests were biological, not product development, so I accepted a postdoctoral research position at Dartmouth College studying lipid biochemistry. It was late in 1962 by then, and that's where and when the DDT issue surfaced as a major distraction and ultimately as a second, unpaid career that carried on for a decade. Its residue in my life remains today.

So what is DDT? It stands for the shortened name dichlorodiphenyltrichloroethane; the full correct name is 1,1,1-trichloro-2,2-bis(p-chlorophenyl) ethane. You don't need to remember that. As the first in a new family of pesticides, DDT had spectacular success during World War II by blocking transmission of several important insect-borne diseases, especially typhus and malaria (Dunlap, 1981). Insects had no prior genetic experience with this chemical, and they were highly susceptible to its toxic action. Man's battles against insect pests would be over, it was thought. DDT was the miracle insecticide. As the war ended in 1945, demand was great, virtually unstoppable, for the use of this material to combat every imaginable insect pest problem. By 1946 the use of DDT was worldwide, and great quantities were being released into the environment.

During the war, DDT had been rushed into service without anything resembling adequate testing for side effects to nontarget organisms. That is entirely understandable, with many earlier wars having been decided not only on the battlefield but also by diseases. DDT proved toxic to insects, its acute toxicity to humans was low, and there was a war to be won. Few voices were raised concerning potential problems for nonhuman organisms.

It was not long after the war before unanticipated problems appeared. Bird mortality and failures at insect control appeared within a few years; the 1950s saw an avalanche of papers describing an assortment of problems with birds, fish, mammals, human health, and ecological disruptions; and Rachel Carson's *Silent Spring* was published in 1962 (Carson, 1962). By 1970 there were many hundreds of scientific papers representing an immense amount of research demonstrating that DDT contamination had become a serious worldwide problem.

Coincident with damage to nontarget organisms and ecosystems was an increasing failure of DDT to perform its intended function of insect pest control. Resistant pest insect populations developed; natural enemies of pests were decimated, leading to population explosions of the pests; and new insect species were elevated to pest status where they were not before. The nightmarish pest problems created in the Canete Valley of Peru by DDT in the early 1950s serve as a classic story, a forerunner of things to come elsewhere, told later in this book.

DDT was doing widespread environmental damage while failing to do its job.

DDT KILLS BIRDS BUT DOESN'T SAVE THE TREES

It was at a party in Hanover, New Hampshire, during December 1962 that Betty Sherrard, a local conservationist and birder, circulated a petition calling for the town not to treat its elm trees with DDT in an attempt to control Dutch elm disease. She described fluttering and dying birds near her property and explained that DDT had killed many birds in earlier years. She was circulating the petition because she did not want to see the bird deaths repeated. I signed it, as did others. When the petition was presented to town officials, they said they had applied DDT carefully according to directions on the label, and that the bird mortality in earlier years had been caused by a nerve disease within the bird population, not by DDT. The petition went nowhere in convincing them to alter their intended DDT spraying program. The response of the town officials was typical. In the early sixties, conservationists were politically weak sisters who were largely ignored.

The American elm tree had become the most popular shade tree since Revolutionary times, and by the mid-20th century its spreading branches lined the streets of countless American towns and villages. Then about 1930 a new disease arrived from Europe, a fungus disease that clogged the vascular system of the elm trees, choking their water- and nutrient-carrying capacity and eventually killing the tree. The disease was described by a Dutch scientist, so it became known as Dutch elm disease, and it can be transmitted from an infected elm to a nearby healthy tree by elm bark beetles that carry the spores of the fungus. Elm trees are majestic and popular, so they were planted side by side in towns, fostering transmission of the disease. When the miracle insecticide DDT came along shortly after World War II, it was employed by thousands of towns in a single spring spraying to kill the bark beetles as they emerge and before they can spread the disease. Who could blame town officials for their efforts to save their beloved elm trees?

Since the elm spray program was to go forward in Hanover in the spring of 1963, four of us, all scientists (except Weber) and birders, decided to conduct a study to see what would happen. Included was my first wife Doris H. Wurster, a pathologist; Walter N. Strickland, a postdoc in biology; and Hans W. Weber, a professor of German. We conducted bird censuses, recording and counting all birds seen or heard before work in the morning and again at lunchtime. We chose study sites in Hanover, to be sprayed with DDT, and compared them with ecologically similar study sites in Norwich, Vermont, a mile west of Hanover across the Connecticut River, which was not sprayed.

The elm trees in Hanover were sprayed with DDT during the nights of April 15–18, 1963. The treatment was done at night when wind would be minimal. The following days we expected to find many dead birds around Hanover. Not so! None! That demonstrates how little we knew about studies, published years earlier, documenting that bird mortality could take days to weeks to become apparent (Hickey & Hunt, 1960; Wallace, 1962). The DDT had to work its way through food chains to the birds. We were not yet familiar with the literature, nor had we read Rachel Carson's *Silent Spring* (Carson, 1962). Carson had little influence on our study, nor did any of us ever meet her before her death in 1964. We were unbiased and empty-headed about DDT as our study began.

Residents in both towns were alerted by radio and newspapers to bring us any dead or dying birds they encountered. Soon people were bringing us sick and dying birds. Within several weeks 151 dead birds, especially American Robins, had been brought to my laboratory and kept in the freezer; 10 dead birds came from Norwich. Lots of the birds came from the Dartmouth College campus and were brought in by students. They had read about our study in the Dartmouth newspaper and they wanted to help.

Many of the birds from Hanover exhibited tremors and convulsions before death, the typical symptoms of DDT poisoning. DDT destabilizes nerves, causing them to fire spontaneously without control, so muscles twitch uncontrollably. We dissected the birds and they were analyzed for DDT content in a laboratory in Wisconsin.

Working up the DDT analyses for the several organs of the birds presented a challenge. I was told I needed to calculate geometric means (averages), but I had no idea how to do that. I had majored in chemistry. It was beyond my slide rule, and mechanical calculators at that time were the size of large typewriters, with many spinning wheels and clanking parts. I was told that Dartmouth College had a computer (the first time I had heard the word), and sure enough, it was in the basement of one of the buildings.

The computer had a room of its own and was the size of an SUV, and there was a person there who knew how to run the thing. The "program" was a deck of cards with millions of holes. I punched the data as more holes into another deck of cards, and both decks of cards were fed into this monster. Out came a vast printout of paper that stretched across the room, and nirvana, there were the geometric means. They were published in our paper in *Ecology* (Wurster DH et al., 1965). I didn't touch another computer for 30 years.

Upon analysis, we discovered that all of the birds exhibiting tremors prior to death on analysis contained lethal concentrations of DDT in their brains (Wurster CF et al., 1965; Wurster DH et al., 1965). No such birds appeared from Norwich; the few dead birds from Norwich had died of other causes.

By this time we had done our homework and caught up with the literature; our findings were quite similar to earlier reports from the Midwest (Wallace, 1959, 1962). It was all right there in *Silent Spring* (Carson, 1962), the book some continue to call "controversial" up to this day. The book was a well-written and accurate review of the literature. Have the critics read it?

The birds had accumulated the DDT by eating contaminated food: ground feeders from soil organisms, and bark and treetop feeders from contaminated insects in the trees. It was especially noteworthy that Myrtle Warblers, treetop feeders that were hundreds of miles to the south when trees were sprayed, were tremoring or dead in Hanover on the same day that they first appeared on our censuses. Food chain contamination had worked quickly and efficiently to kill the birds. We estimated that about 70% of the robins in Hanover had been killed by the DDT spraying. Birds in Norwich were unaffected.

We soon realized that we had rediscovered the stories that earlier authors had already described. Those papers had been published mainly in
conservation-oriented magazines, where public impact might be limited.
We believed the death of the songbirds in Hanover would be a concern
to a wider audience. Accordingly, we published our study in the peer-
reviewed scientific journals *Science* (Wurster CF et al., 1965) and more
fully in *Ecology* (Wurster DH et al., 1965). Scientific credibility was important, since a policy change was among our objectives.

When confronted with the results of this study, Hanover officials
agreed not to use DDT anymore. In 1964 they substituted the far less
destructive insecticide methoxychlor, but because DDT remains in the
soil long after its use, organisms living in soil contaminated by DDT still
killed some birds who fed on them.

By this time we had also learned that DDT is not very effective (nor
is methoxychlor) in controlling the bark beetle that spreads Dutch elm
disease. The beetle breeds and overwinters in dying or recently dead elm
branches, then emerges in spring and flies to nearby healthy elm wood to
feed, thereby spreading the disease. By spraying DDT onto the trees while
they are still dormant, the intent is to kill the beetles as they arrive to feed
in the spring, thus slowing the spread of the disease.

More effective in saving the trees is the practice of "sanitation," the removal and destruction of all breeding material for the beetles—that is, getting rid of dying or recently dead elm branches and even nearby woodpiles.
Sanitation substantially reduces bark beetle numbers in the neighborhood,
and beetles do not arrive from faraway elms. Unlike elms in the forests,
elms in towns were close together and had lined the streets in rows, making
the trees susceptible to the epidemic of Dutch elm disease. This nonchemical sanitation of the elms effectively controls spread of the disease, a technique known and published since the 1930s (Wurster DH et al., 1965).

We had spent two years stopping DDT in one town, while hundreds
of other towns continued to use it, a not especially spectacular performance. Stopping all towns from using DDT on elm trees seemed like a
worthy objective. That objective resurfaced three years later in Michigan
and Wisconsin, but by that time we had a still more ambitious goal.

In September 1965 I moved to Long Island to become Assistant Professor of Biological Sciences at the State University of New York at Stony Brook. Doris and I divorced and she remained in the Dartmouth Medical School. It wasn't long before the DDT issue bubbled back to the surface for me, this time in a rather different capacity. I had met new friends and allies, and the battle against DDT was soon to be escalated.

2

Sue the Bastards on Long Island: The Power of an Idea

During the fall of 1965, a small group of people living on central Long Island, New York, with interests and concerns about a variety of environmental issues had begun to meet monthly in each other's living rooms.* Attendance of 25 to 30 included scientists from Brookhaven National Laboratory and the State University of New York at Stony Brook, in addition to various conservationists and a few high school students. The group called itself by the noneuphonious name of Brookhaven Town Natural Resources Committee, which quickly became BTNRC for obvious reasons.

BTNRC was fascinating and enjoyable, but hardly an organization. There was no office, staff, money, bylaws, elected officers, or any of the

---

* In the early chapters in this book, many passages of my own writing have been taken from *Acorn Days* with encouragement and permission of Marion Lane Rogers, editor and an author, and Environmental Defense Fund, copyright holder (1990).

other ingredients usually present in an organization. It was just a group of people who met occasionally to foster environmental protection policies by our local governments, and we all had other daytime jobs. We discussed various environmental issues—pollution from duck farms, dredging of wetlands, sewage pollution, DDT use on local marshes, dump sites, groundwater protection, wildlife and habitat preservation, and so forth. Meetings usually ended with one-person committees assigned to go do something during the weeks that followed, typically writing a letter to a congressman, a local politician, or a local newspaper. There was no treasury or treasurer, so occasionally we tossed a dollar or two into the middle of the room so that Myra Gelband, one of Art Cooley's dedicated high school students, could send postcards to announce the next meeting.

Attendance was excellent because meetings were fun with good company, good humor, and coffee and donuts at the end. The only feature of this nonorganization was that we had a letterhead printed to give the impression that there was, in fact, such an organization. We needed a bit of puffery to appear greater than we were, for otherwise we feared nobody would listen to us. Everyone seemed to like each other and got along well. An enjoyable social mix is surely a motivational factor that helps explain which groups continue and grow, and which ones stagnate. Another beneficial feature about BTNRC was that we discussed and argued about real environmental problems and how to solve them, which is interesting and can be exciting, and not about getting organized, which is not.

Since some of the BTNRC players become important as our story proceeds, I will introduce them here. BTNRC had no officers, but Arthur P. Cooley became the unelected chairman, since he could run a meeting with Robert's Rules. Art, a biology teacher from Bellport High School, had grown up on Long Island to become an outstanding biologist, ornithologist, field naturalist, and teacher. He had the remarkable ability to arouse people's enthusiasm about environmental topics. It was said that he could get a group of people excited watching a blade of grass grow.

Dennis Puleston was the grand old man of BTNRC. He lent respectability where it was badly needed. Raised in England, Dennis had spent six years wandering the oceans during the 1930s by small sailboat. By 1937 he found himself in Peking (now Beijing) under house arrest by the invading Japanese. A year earlier, Dennis had kept three cockatoos on his sailboat, but when they began to chew on the riggings, he made a gift of them to the Emperor of Japan. The Emperor sent Dennis a thank-you note in reply. When he showed the Japanese a letter from the Emperor of Japan, whom they considered divine, his captors helped him escape into Manchukuo, onto the Trans-Siberian Railway with bread and a sausage, and ultimately back to England.

Dennis settled on Long Island as Technical Information Officer for Brookhaven National Laboratory. He had become a legend in his own time as a world traveler, naturalist, ornithologist, author, artist, photographer, eco-tour leader, and lecturer. There wasn't much that Dennis didn't know or couldn't do. His watercolor of an Osprey is Figure 2.1.

Another major player in BTNRC was George M. Woodwell, senior ecologist at Brookhaven National Laboratory. He had published and lectured widely on a number of ecological topics, including the persistence

OSPREY

Figure 2.1 Watercolor of an Osprey by Dennis Puleston. By permission of Dennis's daughter, Jennifer Clement.

of DDT in forest soils in Maine. George was the most respected scientist in the room, and he was always the strongest advocate for maintaining our scientific integrity.

Robert E. Smolker was an ecologist and ornithologist at the State University of New York at Stony Brook. Dr. Smolker was a walking encyclopedia of biological knowledge, and his side-splitting humor made things fun that otherwise were not. He had an intimate knowledge of local environmental issues and their politics. Bob was a cohesive force among us until his untimely death in 1985.

Anthony S. Taormina was Regional Director for Fish and Game for the New York State Department of Environmental Conservation. Tony seemed to know everything about Long Island's environmental issues and politics, providing much of the insight that contributed to the unusual effectiveness of BTNRC.

I also became a regular attendee of BTNRC gatherings. One frequent topic of conversation was Long Island's Suffolk County Mosquito Control Commission, which since 1947 had liberally applied DDT to local marshes in the name of mosquito control. By the 1960s it was well known to conservationists and many scientists that DDT was destructive to wildlife, but efforts for a decade to reduce or stop the flow of DDT from the Mosquito Commission had been to no avail. Often the Commissioner, Christian T. Williamson, was said to pop a pinch of DDT into his mouth, disarming his audience and leaving environmentalists in disarray.

One BTNRC meeting in April 1966 left me with the assignment of writing a letter to local newspapers, attacking the continued use of DDT by the Commission. I was the logical choice, since I had done research on DDT and its effects on birds at Dartmouth College in New Hampshire. Williamson had often stated that "DDT does not harm animals," so my letter of May 6, 1966, in the *Long Island Press* said, in part:

> If the decline in Long Island wildlife is to be checked, the use of DDT for mosquito control must be curtailed. It is alarming to think that the dissemination of such toxic materials is in the hands of a person who thinks they are harmless.

I certainly didn't realize then that my letter would set in motion an eight-year sequence of events that would change the course of my life, and many other lives, and lead to major changes in national pesticide policies. That night I received a phone call from Victor J. Yannacone jr., a Patchogue, New York, attorney. He informed me that he was representing his wife, Carol, in a lawsuit pending against the Mosquito Commission for causing a fish kill on Yaphank Lake, allegedly by flushing the tank of their DDT spray truck into the lake. Vic wanted to know if I knew anything about the effects of DDT, would I support his lawsuit, and did I know any other scientists who would do likewise. I said yes, yes, and yes.

The next night we were back in Art's living room for a strategy session with Vic Yannacone. "Sue the bastards" was his battle cry. "If you scientists know what you are talking about, a lawsuit ought to get some action," said Vic. We were eager for a way to go after the Mosquito Commission, and Vic sounded as if he had one. He generated a lot of steam and contagious excitement, and several of us came away from that meeting with assignments to gather our scientific reprints, run the photocopy machines, and write affidavits on the destructive effects of DDT on wildlife. With some coaching from Vic, we prepared a substantial package of documentation in less than a week. A few days later Vic rushed off to the State Supreme Court in Riverhead, New York, to file all the papers with Judge D. Ormonde Ritchie, requesting an injunction against the continued use of DDT by the Commission. All this got into the newspapers and gave the Commission lots of bad publicity.

We were stunned when on August 14 the judge issued a temporary injunction against the Commission, enjoining them from further use of DDT on Long Island's marshes. Our lawsuit had stopped DDT use in only a few weeks, where conservationists on Long Island had been after Williamson for a decade to do so, to no avail. *This lawyer with his typewriter (nobody had computers then) seemed to have some strong medicine, and we needed it.* This was very exciting.

It was almost ludicrous when, the day after the injunction was issued, the spray truck went down Vic's street in Patchogue and sprayed his front yard. Vic rushed out with a paper towel, which he immediately brought to

me in Stony Brook for analysis. Yes, it was DDT! Vic charged off to Riverhead, filing a motion claiming contempt of court. The judge was irate, and
the Commission lost more credibility. They claimed they were just emp-
tying their trucks. I guess they didn't know what else to do with it, and the
driver didn't realize he had picked the wrong yard on the wrong street.

Nine years earlier, in 1957, Dr. Robert Cushman Murphy, an eminent
seabird ornithologist who lived in Setauket, Long Island, had gone to
court with other citizens to prevent extensive aerial DDT spraying for
gypsy moths (Carson, 1962). Much of Long Island was sprayed before the
case was considered, rendering it moot. The case seeking a permanent
injunction was appealed all the way to the U.S. Supreme Court, which
refused to hear it. Justice William O. Douglas wrote a vigorous dissent,
but widespread use of DDT continued.

By mid-1966 we were giving some serious thought to the idea that perhaps science and law in court made an interesting combination for environmental protection. Environmental conservation always seemed to be
politically impotent, and we all were amply frustrated at being ignored. We
were tired of losing. Conservationists spent much time complaining to each
other, but not influencing political decision makers. Letters to politicians
received form letters in reply. Not much happened following letters to editors. *A lawsuit seemed to get the attention of our tormentors; maybe it was
like a police whistle in traffic.* It made them read our documents. A competent and correct scientific position, along with legal pleadings, clearly had
some clout. We had grabbed the Mosquito Commission by the tail.

DDT GOES TO TRIAL ON LONG ISLAND

Our case came to trial in Riverhead in late November 1966, continuing
into December, before Judge Jack Stanislaw (Fig. 2.2). Those were the days
when judges had never heard of the word "ecology." The judge had to look
it up in the dictionary before coming into the courtroom. We had a lot
of educating to do if we were to prevail. We were fortified with witnesses,
charts, and reprints. Tony Taormina was our first witness. He said that
DDT was a persistent poison that damaged wildlife, including Ospreys

Figure 2.2 Three EDF founders, Art Cooley, Charlie Wurster, and Dennis Puleston, in 1987, commemorating the 20th anniversary of the founding of EDF and the 21st anniversary of the first trial against DDT in the New York State Supreme Court in Riverhead, Long Island, in November 1966. Photo with permission by T. Charles Erickson.

and Bald Eagles, and that other materials could be used to control mosquitoes. Using my chemistry background, I testified to the properties of DDT, that it persisted in the environment; could move considerable distances from where it was applied; would concentrate in food chains because of its solubility characteristics; and was toxic by several mechanisms to a wide assortment of organisms. I also submitted a "Technical Appendix" containing more than 100 reprints of papers documenting wildlife damage from DDT, which included reference to a book by Robert L. Rudd (1964) that documented more than 300 scientific papers detailing wildlife damage from DDT as early as 1947 in Princeton, New Jersey.

Dennis Puleston had studied Ospreys on Gardiners Island, an essentially uninhabited island between the two forks of extreme eastern Long
Island, since 1954. The island is an ideal nesting habitat for Ospreys,
where they have always been fully protected and the habitat has remained
unchanged. Dennis estimated the island had 300 active nests fledging 600
chicks in 1948; 200 nests fledged 250 chicks in 1954; and by 1965 and 1966
about 75 nests were fledging almost no chicks (Puleston, 1975). This fish-
eating raptor had declined precipitously because nests were not producing chicks, especially because eggs were often breaking in the nest. The
same trends were occurring in nearby Connecticut and New Jersey, and
in all cases, eggs and dead chicks carried high concentrations of DDT.

At this stage, late 1966, we had a general sense that DDT was affecting
the reproduction of predatory birds, including Bald Eagles and Ospreys,
but we did not know how. Studies within the next few years would greatly
clarify this effect and would show that effects on reproduction were having
a devastating effect on their populations. This sublethal effect became
more important than direct mortality as our campaign progressed, but
we did not know that in 1966.

Dennis explained food chains, who eats whom, how a salt marsh works,
all illustrated with seven beautifully drawn food web posters. Figure 2.3
shows one of those charts. Dennis explained how DDT moves through
a food web, becoming more concentrated with each step, reaching the
greatest concentrations in Ospreys at the top. I had done the DDT analyses. As an example, DDT concentration was 0.04 part per million (ppm)
in plankton, 0.16 ppm in filter-feeding shrimp, 0.23 ppm in small fish,
2.07 ppm in larger fish, 13.8 ppm in an Osprey egg, and 75.5 ppm in a
Ring-billed Gull. That's a concentration effect of nearly 1,900-fold from
bottom to top.

George Woodwell explained how DDT travels with circulation patterns of air and water and gets around the world, contaminating most
of the world's organisms and reaching higher levels in carnivores at the
ends of food chains. We had collaborated in research on the accumulation of DDT in the Long Island marshes and the organisms that live in
them. This study was later published in *Science* (Woodwell, Wurster, &

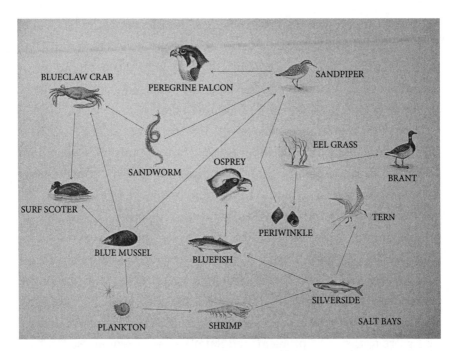

Figure 2.3 One of seven estuarine food web charts presented as evidence in the New York State Supreme Court, Riverhead, Long Island, by Dennis Puleston in *Yannacone vs. Mosquito Commission*, November 1966. By permission of Dennis's daughter, Jennifer Clement.

Isaacson, 1967), while George's testimony (along with Dennis's charts) was later summarized in *Scientific American* (Woodwell, 1967). We were well prepared for this trial, which was a bit like teaching a college course, and the judge was attentive and fair.

The Mosquito Commission did not have a lot to say. They had no biologists on their staff. The New York State Commissioner of Health submitted a "friend of the court" brief stating that "DDT has not been shown to cause irreparable damage to wildlife" and "Most of the evidence concerning DDT and its secondary effects are circumstantial and for the most part based upon a relatively few and insignificant number of observations." The Commission's defense seemed like irrelevant puffery that did not refute any of our well-supported assertions.

The judge must have been impressed with the validity of our case because he kept the temporary injunction in place through the winter and into 1967. The Mosquito Commission got used to not using DDT; they used another insecticide called Sevin. We didn't know it then, but DDT was never to be used again on Long Island. In 1967 it was prohibited by the Suffolk County Board of Supervisors and by the Town of Huntington, and by 1971 DDT was banned in New York State.

We had moved on to greater targets in other states in the DDT issue when unexpectedly, on December 6, 1967, Judge Stanislaw rendered his decision, a year and a half after we had filed the case. He threw us out of court, case dismissed. The Mosquito Commission, the judge ruled, was created by the Suffolk County Legislature and given the charge to control mosquitoes. It was not up to his court to tell them how to do it. Our remedy, to secure a ban on DDT, lay with the legislature, not with the court. His court did not have jurisdiction in our case, he said, nor did we have standing to bring this lawsuit. But the good news, for us at least, was that the court case generated publicity and alerted the public and the legislature to the DDT problem, and Suffolk County would use DDT no more. The judge had kept the temporary injunction in place long enough for all that to happen. *We had won while losing.*

We did a lot of thinking about this strategy, had many discussions about it, and even wrote each other memos on the subject from mid-1966 to mid-1967. We had educated the public through the newspapers, exposed the DDT issue, put an almost immediate halt to the use of DDT by the Commission, and then lost the court case. To a group of frustrated environmentalists, this was clearly an interesting approach: *You could win by winning, or you might get much of what you were after while losing.*

The legal basis for Vic Yannacone's early lawsuits was primarily constitutional. Environmental quality, he argued, is protected by the Ninth Amendment, which says, "The enumeration in the Constitution, of certain rights, shall not be construed to deny or disparage others [such as the right to an uncontaminated environment] retained by the people."

Furthermore, the Constitution guarantees that all citizens shall enjoy equal protection of the laws and, in the Fifth Amendment, that "nor shall any person . . . be deprived of life, liberty, or property, without due process of law." In this case, property includes environmental quality and wildlife. Vic insisted that only these basic constitutional arguments could ultimately prevail. These legal concepts seemed logical enough to the rest of us nonlawyers. Future judges were somewhat less impressed.

There were additional legal problems with our strategy of taking environmental problems to court. Jurisdiction was one of them, but another involved sovereign immunity, which said we could not sue governments, federal, state, or local. Then there was the issue of standing, our right to go to court representing the environment, a nonfinancial cause. The courthouse door, in effect, was locked, and what we were trying to do could not be done.

Our strategy was to keep knocking on that locked courthouse door. Sooner or later, we figured, somebody will forget to lock the door, and one way or another, we would get our foot in the door. Amazingly, that's what eventually happened, but that's years ahead of our story. Much water will flow over the dam before we get to that breakthrough.

RIGGING THE AUDUBON CONVENTION; A WALL STREET LAWYER WAS UNIMPRESSED

During much of 1967 we were casting about for a way to go forward with the DDT issue, but we had no organization, no money, and no forum or mechanism for initiating further actions. Our thinking began to crystallize when Vic Yannacone was invited to address the convention of the National Audubon Society during the last few days of September in Atlantic City, New Jersey. But Vic's ideas soon went far beyond a mere speech to the convention. He wanted to "take over" the convention and to steer Audubon's membership into establishing an "environmental defense fund" to fight Audubon's environmental battles in court. Vic would be legal counsel to such an entity, which would be an arm of National Audubon. In the car on the way to the convention, Vic and Carol Yannacone; Robert Burnap, a staff member of Audubon; and I plotted intricate strategy on how we would "rig" the convention to make all these things happen.

This was all very exciting, to say the least: We had never rigged a convention before. Further strategy sessions at the convention involved one Audubon Director, H. Lewis Batts (from whom Vic had taken a biology course at Kalamazoo College in Michigan), and Audubon's Vice President, Roland C. Clement, among others.

Many hundreds of Audubon members soon filled the convention floor. After Vic's speech, we went into action. To avoid the appearance of collusion we were scattered about the convention floor. I was supposed to go first. Recognized by the chair, I introduced a resolution that DDT was a serious environmental problem and that the National Audubon Society should undertake a major campaign to secure a national ban on its use. Lew Batts, up on the stage with the other directors, seconded the motion, which then passed by unanimous standing vote of the entire membership. We were off to a good start in our little conspiracy.

Bob Burnap then took the microphone to move that the National Audubon Society establish an environmental defense fund with money from Audubon's Rachel Carson Memorial Fund, that this environmental defense fund would fight environmental battles in court, and that the first case would be to seek a national ban on the use of DDT. Roland Clement rose to point out that he had been urging such action by National Audubon for some time. The membership, sensing some action cooking instead of the usual dull business meeting, again rose in unanimous support of the motion.

Here was the Audubon membership directing management to go to court against DDT. Gene W. Setzer, chairman of Audubon's Board of Directors and chair of the meeting, was beginning to look uncomfortable, as were most directors, sitting on the stage facing the audience and unable to communicate with each other. Clearly, things were beginning to get out of hand. Filing lawsuits against corporate interests, some represented on Audubon's board, was probably not an appealing course of action to Audubon management. In those days, filing lawsuits was just this side of throwing bombs: Respectable people didn't do such things.

Quick thinking by Setzer saved the day (and spoiled ours). He suggested that this motion should be referred to Donald C. Hays, Audubon's general counsel, to be considered for action. The membership acquiesced. Our little fire had been transferred to the back burner. We assumed Hays would sit on the idea. Two days later Vic and I visited Hays in his Wall Street law office. Our assumption was right: Hays was not about to let the National Audubon Society get into the business of filing lawsuits about the landscape. The idea would stay on the back burner, and we were not going to have an environmental defense fund within National Audubon that would take environmental issues to court. If it was going to happen, *we would have to do it ourselves.*

THE ENVIRONMENTAL DEFENSE FUND IS BORN

Never doubt that a small group of thoughtful committed citizens can change the world. Indeed, it is the only thing that has. Margaret Mead.

The stage was set for the incorporation of an independent environmental defense fund. The 10 signers of the certificate were the ringleaders of the Long Island DDT suit, as well as the "troublemakers" from the Audubon convention. Lew Batts, who hated to fly, had not yet taken his train back to Michigan and was still in town. Roland Clement was an obvious signatory, but he declined the opportunity; his employer, National Audubon, might not take kindly to its vice president joining this loose cannon that was going to fire lawsuits about, a disreputable thing to do in those days. He accepted a trusteeship a few years later as the strategy began to succeed and gain respectability.

And so it was that on Friday, October 6, 1967, nine men and a woman gathered in that conference room at Brookhaven National Laboratory on eastern Long Island to sign the certificate of incorporation of the Environmental Defense Fund (EDF). Signers included the conspirators from the Audubon convention; Vic Yannacone and his wife, Carol; Lew Batts; Bob Burnap; and myself. Then there were the BTNRC troublemakers Dennis Puleston, George Woodwell, Bob Smolker, Art Cooley, and Tony Taormina (see Appendix 1).

EDF was born from the frustration of a group of environmentalists unable to move the system, to make it respond, to force environmental protection; the frustration of losing environmental battles, often irreversibly; the deep conviction that we had an idea that was going to work because we had seen it work; and the final frustration that even our friends, the National Audubon Society, were not going to buy the idea. *EDF would have to go it alone.*

But if this turned out to be an important event in the history of the environmental movement in America, it was anything but obvious to those present. There had even been confusion about the time, place, and purpose of the meeting. Somebody said it was "Yannacone's meeting," while another thought Bob Burnap had called it. Confusion was probably part of the plan. Vic, the only lawyer in the group, was a master at generating confusion among his opponents, and he sometimes kept the edge on this talent by practicing it on his friends as well. But there must have been a plan, for Vic had arrived with the typed document, and all that was needed were the signatures.

Almost all! There was the small matter of $37 to register the organization in Albany as a not-for-profit corporation in the State of New York. The fee could, of course, be paid by the treasurer of the new Board of Trustees from EDF's funds, but there was no treasurer and there were no funds—not a dime. After a minor hassle marked by a lack of volunteers, Bob Burnap agreed to pay the fee from private conservation funds he had secured from a "sugar daddy" in Connecticut. For that contribution, we elected Bob Burnap to be EDF's first treasurer.

None of those present could know, or even imagine, that in less than a decade EDF was to explode into a highly sophisticated and effective national environmental organization with five offices housing more than 50 full-time staff members, a budget in the millions, numerous legal actions across the land, and a public membership of more than 50,000. Forty years later all those numbers were up by five- to ten-fold, with additional offices in China and Mexico, and climate change, the world's greatest environmental challenge, was the major issue on EDF's front burner.

The EDF of October 6, 1967, bore no resemblance to that organization of the future, nor did it for several years thereafter. There were no offices, no staff, no money, no membership, no bylaws, not even a letterhead—in short, none of the ingredients that fit the definition of an organization.

But the new EDF had one asset—an idea—in its collective mind. EDF would marry science and law to defend the environment in the courts. Courts had not been used for environmental protection, and the new trustees shared the idea that there was an empty niche on the American scene for an organization designed to litigate on behalf of environmental quality. Influencing the legislative or executive branches of government often requires large numbers of votes or dollars, and we had neither. With those approaches, many people had to be convinced. The judiciary seemed less influenced by those traditional pressures, and only one or a few judges needed convincing. Perhaps judges would more readily listen to a handful of scientists if their plea was valid.

It was an idea whose time was about to arrive, but this, too, was invisible in 1967. A decade later, "environmental law" had become a recognized subject taught in law schools and a strategy employed routinely by numerous environmental organizations, including some that were suspicious of it in 1967. Before then, the American civil rights movement had successfully employed the courts to attain many objectives. Indeed, the Environmental Defense Fund drew its name from the Legal Defense Fund of the NAACP; Bob Burnap had coined the name. Vic Yannacone must have liked it, because he typed it into his incorporation certificate, but we never discussed or voted on it. We immediately and automatically called it EDF; the full name was too long. *EDF was it—no resolution, no vote, just EDF.*

Considering EDF's nonexistence as an organization, it was presumptuous—arrogant might be the better word—that EDF was designed from the very beginning as a national organization. It was not a local organization that became national with maturity, but rather an utterly immature national entity. The purpose seemed narrow in scope. Despite verbiage about "scientific, educational and charitable purposes" and "to encourage and support the conservation of the natural resources of the United States of America" in EDF's certificate, its every intent was to "sue the bastards" (Vic's slogan). EDF was born of the frustrations of those 10 people with the American system as it then existed for protecting environmental values. All were weary of writing letters to congressmen only to receive reassuring form letters in reply. Dragging the enemy into court seemed just about right, surely exciting, and perhaps even fun.

3

Proceed with Caution, then Sue the Bastards in Michigan

At that first meeting on October 6, 1967, the new trustees of EDF had voted to "proceed with caution," given the precarious position of this essentially nonorganization with no assets. It was an easy motion and it passed unanimously, but before long caution was thrown to the winds when Lew Batts described an imminent planned application of the insecticide dieldrin in western Michigan. Intended to eradicate an alleged infestation of Japanese beetles, dieldrin was to be applied to 3,000 acres in Berrien County near Lake Michigan by the Michigan and United States Departments of Agriculture. Lew wanted EDF to stop them.

We already knew something about dieldrin, a chlorinated hydrocarbon relative of DDT and an environmentally destructive material, more acutely (immediately) toxic than DDT. We knew it would kill birds and mammals and could damage fish. Furthermore, Lew Batts was connected with a Michigan foundation that had more money, but less arrogance, than we did. EDF was designed to litigate, and Batts's organization certainly was not. He guaranteed the assembled new trustees of EDF that if we would tackle the Michigan Department of Agriculture (MDA) in court to block the dieldrin application, he would support the effort with $10,000.

The fat was in the fire! EDF's trustees voted to *cautiously* sue the Michigan Department of Agriculture, and anybody else if necessary, to prevent the dieldrin treatment. Furthermore, several communities within the Lake Michigan watershed in western Michigan were using DDT in an attempt to control Dutch elm disease, a futile exercise with which we were very familiar (Wurster DH et al., 1965). With both of these destructive chemicals contaminating the fish, it would be difficult to separate the effects of each chemical from the other. So we decided to sue not only MDA in connection with its proposed dieldrin application, but we would add as defendants nine cities in western Michigan within the Lake Michigan watershed that were using DDT (Fremont, Muskegon, Greenville, Rockford, Lansing, East Lansing, East Grand Rapids, Holland, and Spring Lake).

The environmental hazards of dieldrin were well documented in the literature (Carson, 1962; Wurster, 1971). Figure 3.1 shows a pile of dead birds following a dieldrin application. The fear was that the dieldrin would not only damage bird and mammal populations, but that it would also run off the land into Lake Michigan and damage fish, especially the introduced and highly successful Coho salmon fishery, so popular with fishermen. There was also evidence that DDT was already present in the salmon and might be inhibiting their reproduction (Macek, 1968a, 1968b).

In our complaint we would treat dieldrin and DDT together, as though they were one hyphenated word. Our complaint not only detailed the harmful effects of dieldrin and DDT to wildlife but also offered harmless yet effective alternative pest control strategies. Sanitation of elm trees, in fact, saves the elm trees where DDT does not (Wurster DH et al., 1965). We had already learned that *battles are easier to win when alternatives are offered.* In later actions on DDT, many of our key witnesses were top experts in the developing field of integrated pest management. Presentation of alternatives has been a consistent theme throughout EDF's history.

Figure 3.1 Dead birds following dieldrin application, Sheldon, Illinois, 1954. Offered into evidence, Federal Court, Grand Rapids, Michigan, October 30, 1967.

The applications were scheduled to begin about November 1, so there was no time to lose. It took two weeks to get our papers together, and on Friday, October 20, Vic Yannacone, Bob Burnap, and I flew to Kalamazoo, where we joined Lew Batts. Lew had arranged for us to use the law office of his attorney, Edward P. Thompson, in Kalamazoo for the weekend to finish our complaint. A feverish weekend it turned out to be: Day and night we worked on the complaint, sleeping a few hours in the office on Friday, Saturday, and Sunday nights.

One incident stands out in my memory. The dieldrin application was to be jointly funded by MDA and the U.S. Department of Agriculture (USDA). For this reason, Vic seemed convinced that USDA also had to be a named defendant—but he didn't want to name USDA. He was clearly uncomfortable by that prospect, and he had been fretting all day Saturday and Saturday night to find a way not to include USDA. It was 3 a.m. on Sunday and Vic was still fretting. He disappeared into the bathroom and did not emerge for an hour. I was sleeping on the floor.

Finally he reappeared. Without a word he took a razor and carved a hole in the front page of the complaint; he had cut USDA out of the suit. I never understood it, but in the bathroom he had figured out a way to avoid suing USDA. (Just imagine! There were no computers in those days. To make changes, you used Wite-Out, erased and risked making a hole in the paper, carved out the offending part, or retyped the whole page.)

On Monday morning, October 23, 1967, the complaint was finished, and we marched into Federal District Court for the Western District of Michigan in Grand Rapids to file it. We joined as defendants the MDA and nine cities in western Michigan, seeking to stop the use of dieldrin and DDT as well. Only 17 days after incorporating, EDF had gone to court for the first time in its own name.

Along with many Michigan residents, we also sent a telegram to Governor George Romney asking him to stop the dieldrin application. A great deal of publicity followed, not only throughout Michigan but nationally as well, and the *Detroit Free Press* called us "a powerful group of scientists." We certainly didn't feel very powerful, and the group was rather small, but we appreciated the importance of public opinion favoring our cause, and inflating ourselves a bit was part of the strategy. By this time Bob Smolker had created his "Mystique Committee," of which he was chairman and sole member, and it seemed to be working. Puffery was his product. One-person committees don't have meetings.

Whereas EDF was in court against MDA, EDF had the active support of the Michigan Department of Natural Resources, the department responsible for the Coho salmon success. The director, Ralph A. MacMullan, and his deputy in charge of pesticides, C. Ted Black, were especially helpful with information, ideas, and political support. They did not want to see DDT damage their salmon fishery. EDF also enjoyed the support of several Michigan environmental organizations, which provided hospitality, contacts, and funds for meals and airfares. Local support was vital to EDF, which otherwise might have been viewed as a New York corporation (which it was) invading foreign turf.

THE SUIT PARTLY UNRAVELS; ANOTHER WIN WHILE LOSING

After all our masterful strategy and effort of wrapping dieldrin and DDT into the same package, the judge immediately split the action in two. There was the DDT action against the nine cities, and there was the dieldrin case against MDA, with the scheduled dieldrin application only days away. After reading our documents, the court issued a temporary restraining order blocking MDA from applying the dieldrin. Again litigation had stopped an environmentally destructive action by an agency, but our jubilation was short-lived.

A week later we were granted a two-day hearing in which George Woodwell, Bob Smolker, Roland Clement, Lew Batts, and I, along with Ralph MacMullan and several MDA officials, testified before the court. Among other points of interest, the "infestation" of Japanese beetles consisted of about one beetle per acre, but our limited time allowed only a partial presentation of the case. Then, just like what happened on Long Island, our case was kicked out of court, dismissed on the grounds that we lacked standing to sue a state agency. The standing issue was clearly a large problem. The restraining order was dissolved. MDA was now free to apply the dieldrin. Gloom prevailed in our camp.

But a new and unknown ally shortly intervened on our behalf. It was November, and snow began to fall in western Michigan, again blocking the dieldrin application. This time the restraint was permanent: Winter had arrived for good. We clearly had friends in high places, and again *we had won while losing.*

Litigation was a wondrous tool! We did not know then that MDA would finally succeed with a reduced dieldrin application a year later, but by then we had other fish to fry in Wisconsin in the fall of 1968.

THE CITIES CAVE IN, FORGET TO KICK US OUT

Meanwhile, back at Federal Court in Grand Rapids, the DDT action against the cities that planned to use DDT on elms was still pending.

The cities had decided to fight us with PhDs who would refute our evidence. We anticipated a trial, but months passed: Apparently the cities were having trouble finding such PhDs. One by one the cities threw in the towel and accepted stipulations that they would stop using DDT in their efforts to control Dutch elm disease. It wasn't long before we had court orders requiring all nine cities to stop the use of DDT. Although we got the distinct impression that the judge would have granted a motion to dismiss, the cities never made the motion. *We had won without a fight, surviving in court because someone forgot to kick us out.*

This victory must have gone to our heads, for we charged forward by adding all the other cities in the state of Michigan—43 of them, including Detroit—to the front page of our complaint. This move seemed especially expeditious, since a new front page was the only new document needed.

History repeated itself. The added cities also were unable to refute the evidence and, one by one, they accepted stipulations against any further use of DDT. Within a year, 51 of 52 cities had thrown in the towel. The one holdout, a northern suburb of Detroit, stubbornly refused to concede. But 51 out of 52 was good enough for us; we pursued that town no further. The use of DDT for attempted control of Dutch elm disease had been stopped in Michigan. It was a funny mix of getting kicked out of court for having no legal right to be there, winning while losing, winning by default, good science, and divine intervention.

The Michigan State Legislature banned the sale of DDT in Michigan on April 17, 1969. We told ourselves we had something to do with that.

EVOLVING A MASTER PLAN

By the spring of 1968 we had evolved a rather clear vision of where we thought we were headed. Litigation was a powerful tool for environmental protection, and there appeared to be an opportunity for its implementation in the United States. It promised to put some teeth into the heretofore weak environmental movement. But there were plenty of problems. One was sovereign immunity: "You can't sue the Crown" because "the

King can do no wrong." As America inherited British law, citizens cannot bring suit against their own government, and in the eyes of the law, EDF is equivalent to a citizen.

A different and potentially more serious problem, which we did not fully understand at the time, was the issue of "standing" to sue, which is different from sovereign immunity. To have standing to enforce a particular law, a citizen, such as EDF, must have a direct and tangible stake or interest in the matter before being allowed to come before the court. Was EDF destined to continue to win battles, or perhaps to start losing them again, while being thrown out of court because of "sovereign immunity" or "standing" barriers?

We had an optimistic plan for these problems. If the courthouse door was always locked, and we kept pounding on the door anyway, somewhere, some day, somebody would forget to lock it (as happened with the cities in Michigan), or someone would open it for us, and we would gain entrance to the court. That precedent would make it easier the next time—and the next. We would break down the legal barriers by trying again and again. Meanwhile, perhaps our luck would hold and we would continue to win some environmental gains even while losing the legal proceedings.

Joseph L. Sax, professor of law at the University of Michigan and the leading advocate for developing environmental law, gave us invaluable advice for the pursuit of those objectives. He explained how government regulatory agencies had become captive to the industries they are supposed to regulate and that laws already on the books to protect environmental values were not being enforced. Tackling this problem through the courts then ran into sovereign immunity and the standing issues, meaning that citizens (EDF) cannot sue government agencies. A persistent program of litigation to force agencies to enforce laws appeared to be the best way to break down these legal barriers and to allow citizens to participate in decisions that gravely affect their lives.

To follow that strategy, we had to have a scientifically impeccable case, one so convincing and scientifically strong that we would not lose on the scientific merits. If we were going to lose on law, we had to win on science if we were to have any chance of prevailing. We had to convince judges of the merits of the case, giving them reasons to find excuses for opening the courthouse door to let us in. We figured that the DDT issue was the strongest scientific case we were likely to find. DDT would break down the courthouse door so that other, less powerful environmental cases could enter. The neighboring state of Wisconsin was soon to offer just such a possibility.

A major ingredient of our strategy must not be overlooked. By going through the court system, a small number of judges might make important decisions on the merits, free of political or financial pressures. We were avoiding state legislatures and Congress, where such pressures prevail, and where large numbers of legislators and the general public would need to be convinced of the validity of our actions. Our approach through the courts would attempt to force government agencies to take effective regulatory actions under existing law. *We were not going through legislatures or Congress seeking new laws.* Others might wonder whether the DDT issue would have been resolved as it was, had we attempted to go through Congress. That seemed most unlikely, then or now.

4

On to Wisconsin, the Dairy State

EDF was only a few months old with the DDT agitation still under way in Michigan when a phone call arrived from Lorrie Otto, an environmental leader in Milwaukee, Wisconsin (Fig. 4.1). Lorrie explained that DDT was to be used for attempted Dutch elm disease control in Milwaukee, and she wanted EDF to come to Wisconsin and stop it. She had been reading about our fireworks on Long Island and in Michigan and figured some of that might be good for Wisconsin.

Other considerations were also favorable about Wisconsin. The use of DDT leads to contamination of meat, eggs, milk, and other dairy products, and the dairy industry is large and important in Wisconsin. Agriculture was therefore split in that state: One component wanted to use DDT on crops and trees, while the dairy industry wished to avoid DDT contamination of its products. Agriculture would not present a unified force on behalf of DDT, as it would in most states. The environmental movement was unusually strong in Wisconsin, and Lorrie Otto pledged

Figure 4.1 Lorrie Otto, the human catalyst who made the Madison
DDT hearings happen. Photo about 2006 by permission of her
daughter, Patricia Otto.

to EDF the statewide support of the Citizens Natural Resources Association (CNRA). CNRA would raise money and provide all additional logis-
tic support that might be necessary during whatever proceedings might
develop. If EDF was going to make a difference on the national scene, this
seemed like an invitation not to be refused.

So here was an opportunity to stir up trouble again with an increase
in complexity and confusion. On October 2, 1968, EDF filed a lawsuit in
Federal Court in Milwaukee that was quite similar to the one filed nearly
a year earlier in Michigan. It sought to block the Michigan Department of
Agriculture (MDA) from applying dieldrin in Michigan against Japanese
beetles (same case, same application as postponed last year) and to stop
Milwaukee from spraying DDT on its elm trees. The allegations were the
same: both DDT and dieldrin would kill nontarget birds and mammals,
and both would threaten reproduction of the Coho salmon in Lake Michigan hatcheries. Almost before we knew it, the dieldrin case was moved
back to the same court in Michigan where it had been dismissed the year
before. Another injunction was issued, which then expired, and MDA

was free to apply the dieldrin. We lost, but postponed the dieldrin application for a year, then lost the case again by losing, and MDA went ahead with a smaller dieldrin application.

Meanwhile, our action against DDT use in Milwaukee took a dramatic turn. On October 18 the city agreed not to use DDT on its elms, and EDF agreed to withdraw its suit against the city (we might have been thrown out anyway). Brokering this deal was the hearing examiner for the Wisconsin Department of Natural Resources, Maurice Van Susteren, who explained how we could take another course of action, a really intriguing aspect of the Wisconsin situation.

He explained how Wisconsin has a water pollution law that provides for a hearing to determine whether a substance under challenge meets the definition of a pollutant of the waters of the state. If declared a water pollutant, the substance under challenge can be barred from further use within Wisconsin. If Wisconsin citizens and EDF were to file a legal petition challenging DDT as a water pollutant, the Wisconsin Department of Natural Resources would hold the hearing and make the decision. An important consideration was that no court standing issue was involved: The hearing would take place before a state agency, not a court, with no legal challenges on standing or jurisdiction. Here was an opportunity for EDF to present the full case against DDT.

Meanwhile, EDF had not gotten organized. All of our attention and efforts were devoted to developing strategies to fight the DDT battle. Dennis Puleston was still chairman, Bob Burnap was treasurer with no treasury, there was still no staff, office, bylaws, tax exemption, or five-year plan, which all organizations are supposed to have. At that point, EDF was weak on organization and money, but getting the science right was its strong hand. George Woodwell chaired a Scientists Advisory Committee with a growing list of scientists in many fields who supplied expertise and testimony on request, all without fee. Scientific accuracy and credibility has long been EDF's strength. We would go forward with a powerful scientific position, and with weak but promising law.

DDT GOES TO TRIAL IN MADISON, WISCONSIN

On October 28, 1968, the petition for "declaratory judgment" by the Wisconsin Department of Natural Resources as to whether DDT is or is not a pollutant of the waters of Wisconsin was filed. Six Wisconsin citizens, CNRA, and the Izaak Walton League were the petitioners under Wisconsin's unique water pollution law. EDF was invited by these environmental organizations to handle both the legal and scientific aspects of the case. The resultant hearing was set to begin on December 2, with Van Susteren presiding as hearing examiner.

Our strategy in preparing our case was to depict the pesticide industry as narrow-minded specialists knowing only that DDT kills insects, all insects, and that it does not kill people outright. We would describe environmental scientists as a coalition of scientists from many diverse disciplines in constant communication with one another. Such a crude poison as DDT would cause widespread damage to wildlife and would disrupt whole ecosystems, including agricultural ecosystems. We would show that DDT is inherently an uncontrollable substance once let out of the can and that it would contaminate organisms far removed in time and space from the site of application. We would also demonstrate that DDT represents a crude and often ineffective approach to insect control, killing the natural enemies of the pests and often making pest problems worse. Instead, we would offer evidence that integrated pest management employing chemical and biological techniques would lead to better insect control and fewer environmental problems, all at lower cost and using no DDT. We would not emphasize human health hazards, although we intended to show that DDT had never been adequately tested.

We were offered a remarkable amount of hospitality and assistance in Madison. Lorrie Otto seemed to know everybody in Madison, and we were housed in her friends' homes, breakfasts included. We had other meals in restaurants and Fred Ott, CNRA's enthusiastic fundraiser, always picked up the tab. University of Wisconsin (UW) faculty became consultants and witnesses. There were typing pools, a messenger service, and library searches all available to us when we needed them. When an important reference surfaced in the hearing, UW students rushed to the library and came back with a copy within an hour. Evenings were filled with intensive strategy sessions with witnesses for the days to follow. All of this was on a volunteer basis. We didn't pay for anything—which was good, since we didn't have any money.

Much of this was arranged by Lorrie. She had become a national figure on natural landscaping. She wanted her yard in Milwaukee to have entirely native vegetation and wildflowers. Meanwhile, Milwaukee had an ordinance requiring residents to mow their lawns. Lorrie refused, which brought her into direct confrontation with "the lawn police." After a long struggle, Lorrie prevailed, created a beautiful wild garden in Milwaukee, and became an environmental heroine. Lorrie was well connected to CNRA, which turned out to be a wonderful ally in the proceedings that followed. CNRA was a well-organized small army of volunteers, which included Fred Ott, as fundraiser. The amazing grassroots, volunteer efforts by the people of Madison are thoroughly documented by Bill Berry in his recent book, *Banning DDT: How Citizen Activists in Wisconsin Led the Way* (Berry, 2014).

The hearing opened on December 2, 1968, with considerable fanfare under the very tall dome of the State Capitol Building in Madison (Wurster, 1969a) (Fig. 4.2). There were plenty of observers and reporters. Gaylord Nelson, popular U.S. Senator from Wisconsin and a well-known, long-term advocate of environmental protection, was our first witness. "In only one generation, [DDT has] contaminated the atmosphere, the sea, the lakes, and the streams and infiltrated the tissues of most of the world's creatures, from reindeer in Alaska to penguins in the Antarctic, including man himself," said the senator.

Vic Yannacone then stunned us all by calling Louis McLean, opposing attorney representing the Task Force for DDT of the National Agricultural Chemicals Association, as an adverse witness. He wanted to get some of McLean's statements from his *BioScience* paper (McLean, 1967) into the record. "The primary contaminants of the atmosphere are natural substances such as dusts, pollen, viruses, and bacteria," said McLean. Most of his paper blames pollution on natural substances.

In this overall view of the State Assembly chambers to-day, witnesses and interested citizens await testimony in the first of a series of public hearings into the use of the pesticide DDT in Wisconsin. Scientists, conservationists and agri-business representatives from throughout the state and nation are scheduled to appear. See story on Page 1. (Staff Photo by Skip Heine)

Figure 4.2 Day One of the DDT hearings, December 2, 1968, under the dome of the State Capitol Building, Madison, Wisconsin. With permission from the *Capital Times* archive.

(Fast-forward 40 years, and some in the fossil fuel industry are telling us that global warming is a natural phenomenon caused by volcanoes, sunspots, natural cycles, and our exhales of carbon dioxide. Sound familiar?) McLean then continued:

Long ago it became apparent that the pesticide controversy was led by two types of critics—purposeful and compulsive. The purposeful include those who use the controversy to sell natural foods at unnatural prices, to give color to their books, writings, and statements, to gain notoriety, or in any way to profit from the controversy. The compulsive were described by Sigmund Freud . . . as neurotics, driven by primitive, subconscious fears to the point that they see more reality in what they imagine than in fact. . . . The anti-pesticide people, in almost every instance hold numerous beliefs in nutritional . . . and medical quackery and . . . they oppose public health programs. . . . While presenting a holier-than-thou attitude, they are actually preoccupied with the subject of sexual potency to such an extent that sex is never a subject of jest. The anti-pesticide leader, as distinguished from the fair-minded person who is merely misinformed about pesticides, can almost always be identified by the numerous variant views he holds about regular foods, chlorination and fluoridation of water, vaccination, public health programs, animal experimentations, food additives, medicine, science, and the business community, or by his insistence that insecticides should be mistermed "biocides."

Testifying unexpectedly and without preparation about his earlier writings seemed to unnerve McLean, and he never fully regained his composure. Then came another adverse witness, Ellsworth H. Fisher, a UW professor and entomologist and an outspoken DDT advocate. The intent was to put into the record a rough outline of the opposition's case to come. "You can't shoot at empty air," said Yannacone.

After a few days the hearing was moved to a smaller room near the Capitol in Madison. The initial EDF approach was to describe the "Wisconsin Regional Ecosystem," with emphasis on its interconnections and interrelationships, prior to presenting any testimony about DDT and its role when injected into that ecosystem. Hugh H. Iltis, a UW botany professor, described these plant–animal–environment relationships in detail. Orie L. Loucks, a UW botany professor and systems analyst specializing in sustainable ecosystems, followed with testimony about air currents, weather fronts, long-distance pollen dispersal, and nutrient cycling, again emphasizing how events in one region could have distant effects. Hugh and Orie both showed the inseparability and interrelationships between agricultural regions and the overall ecosystem of which they are but a part. They thus set the stage for the final systems analysis summation on the last day of the hearings six months later.

With these ecological relationships established, I took the stand to describe the physical, chemical, and biological properties of DDT (Wurster, 1969b). I testified that DDT combines in a single molecule four properties that cause it to be a serious environmental problem. DDT and its metabolite DDE are very stable and persistent chemicals, remaining in the environment for many years. Despite very low vapor pressure and water solubility, DDT is mobile by several mechanisms, moving about the world within currents of air and water to regions remote from the original application site.

Being essentially insoluble in water but soluble in lipids (fats), DDT and DDE partition into and are accumulated by living organisms from the inorganic environment. These chemicals have a broad spectrum of biological activity and toxicity by a number of mechanisms, often damaging the animals they contaminate. DDT and its metabolites not only enter food chains from the inorganic environment, but they are increasingly concentrated toward the top of food chains, thereby posing a particular threat to carnivores (Woodwell, 1967; Woodwell, Wurster, & Isaacson, 1967). These materials pose a danger to nontarget organisms that is almost unique among major pollutants.

My direct testimony took about one hour but was followed by nearly three days of wide-ranging cross-examination by McLean. Many aspects of the pesticide controversy were probed, as well as my background and that of EDF. McLean probably decided that I was the main troublemaker, so he would demonstrate that I was some kind of a nut. It didn't work because I knew most of the main elements of the testimony to come. I wasn't qualified to get into all those topics, but under the rules of evidence in cross-examination, McLean asked the questions, so he was stuck with the answers. He nervously wiped his sweaty palms with a handkerchief the whole time. I was less than the picture of composure myself. On one of those evenings, while we were having dinner in a restaurant, I looked up and was startled to see myself on the *CBS Evening News with Walter Cronkite.*

All of the expected arguments that were used 10 to 20 years earlier against scientists who first reported environmental degradation by DDT were raised, but it was surprising to hear almost none that were new. EDF soon learned that the Task Force for DDT was not well prepared with regard to the current scientific literature on DDT.

Early in the trial, there were repeated attempts to equate DDT with pesticides in general to create the impression that EDF was an "anti-pesticide" organization. References to "pesticides" were invariably met by objections from Vic Yannacone. These were usually sustained by Van Susteren, who reminded McLean to confine his questioning to "DDT," not "pesticides."

BIRDS DIE, NESTS FAIL

In its presentations, EDF emphasized those effects of DDT that affect a whole species and are widespread, even worldwide, in magnitude rather than the more local, although more spectacular, fish or bird kills. The use of DDT for attempted control of Dutch elm disease, however, had long been of special concern in the Midwest, and the dramatic bird mortality caused by this usage was described by Dr. George J. Wallace, professor of zoology from Michigan State University (Wallace, 1959), and Dr. Joseph J. Hickey (Fig. 4.3), wildlife ecologist and eminent ornithologist from the UW, both among the early scientists reporting such damage (Hickey & Hunt, 1960). William Gusey of Shell Chemical Company and a later DDT Task Force witness said that winter spraying of elms, before birds are in the area, eliminates robin mortality, but Wallace contradicted this by telling how robins continued to die for years after the last DDT spraying because they ate contaminated earthworms. We had similar results in Hanover, New Hampshire (Wurster CF et al., 1965).

Wallace described the tremors he observed in birds dying of DDT poisoning. He was followed on the stand by Dr. Alan B. Steinbach, neurophysiologist from Albert Einstein College of Medicine, who gave a comprehensive description of the mechanisms of nerve transmission and their disturbance by DDT. He discussed the Hodgkin–Huxley equation and the effects of various toxins on transmission of the nerve impulse,

Figure 4.3 Joe Hickey, eminent UW ornithologist.
With permission of his daughter, Susan Nehls.

indicating that the effects of DDT are irreversible as compared with these other toxins. "The known mechanism of action of DDT on nerves . . . can account for . . . the observations outlined by Dr. Wallace, both today and in his earlier papers," Steinbach concluded.

Dr. Robert L. Rudd, professor of zoology at the University of California at Davis, described the earliest reported case of biological concentration at Clear Lake, California, about 100 miles north of San Francisco (Rudd, 1964). The lake is a favorite among fishermen but is also plagued by the annoying "Clear Lake gnat." In an attempt to control the larvae of the insect, 14 parts per billion (ppb) of DDD, the somewhat less toxic metabolite of DDT, to protect the fish, were added to the lake in 1949. It worked; fishermen were gleeful. Two more doses of DDD at 20 ppb were added to the lake in 1954 and 1957. Results with the gnat were somewhat less spectacular; resistance to the DDD was developing.

Clear Lake also had a population of 1,000 pairs of Western Grebes before the DDD additions, but during the 1950s they were disappearing. By 1954 many dead grebes were found, but no disease was present. About 15 to 20 pairs were still present in 1958, but they failed to breed. Almost as an afterthought to explain the puzzle, several dead grebes were analyzed for DDD, revealing the astounding level of 1,600 parts per million (ppm) in their fat. That concentration was 80,000 times greater than the concentration originally added to the lake. DDD additions were stopped.

DDD had been concentrated from the water by plankton, then fish, and then the fish-eating grebes. It was the first documented case of biological concentration, a now-familiar pattern for the buildup of fat-soluble stable chemicals, including DDT, DDE, polychlorinated biphenyls (PCBs), phthalate esters from plastics, flame retardants, and many other chemicals within food webs. It explains why so many carnivorous birds at the ends of long food chains have gotten into serious trouble. The solubility properties of the chemicals mean that the fish and birds excrete the metabolic remains of their foods but retain the DDT or other such chemicals, only to be passed to the next higher level in the food chain. The higher in the food chain the animal is, the more DDT will be found in its body.

During the early weeks of the hearing there was objection by McLean to the term "biocide" when used by several scientists in reference to DDT. Included among the pro-DDT witnesses, however, was Taft Pierce of the Orkin Exterminating Company, who testified that DDT is especially useful for killing mice and bats. Pierce said that without DDT, he would have to use "poison" more frequently to control mice. We didn't bother to point out that mice and bats are mammals, not insects, proving that DDT is a biocide as well as an insecticide. *The DDT defenders had shot themselves in the foot.*

The almost ubiquitous distribution of DDT residues in the world environment was the subject of testimony by many scientists. Dr. Robert W. Risebrough (Fig. 4.4), molecular biologist (PhD from Harvard) from the Institute of Marine Resources, University of California at Berkeley,

Figure 4.4 Bob Risebrough and Vic Yannacone discussing strategy during DDT hearings, Madison, Wisconsin, December 1968. With permission of the Wisconsin Historical Society, Image 100,409.

called DDT and its metabolites "the most abundant synthetic pollutant in the global ecosystem" (Risebrough, 1968). He described their dispersal by currents of air and water, and his discovery of these materials in the air over Barbados (Risebrough et al., 1968). His analyses showed oceanic fish and birds from various parts of the Pacific to be contaminated, often to "very high levels." Since Lake Michigan abuts both Wisconsin and Michigan, Hickey gave analytical data on residues of DDE (a DDT metabolite) in the Lake Michigan ecosystem, showing exceptional contamination of many organisms (Hickey, Keith, & Coon, 1966). "Lake Michigan, in spite of its size, is one of the most polluted (with DDE) lakes in the world," said Joe Hickey. The DDT metabolite DDE becomes an important part of this story, so the breakdown pathways for DDT are shown in Figure 4.5.

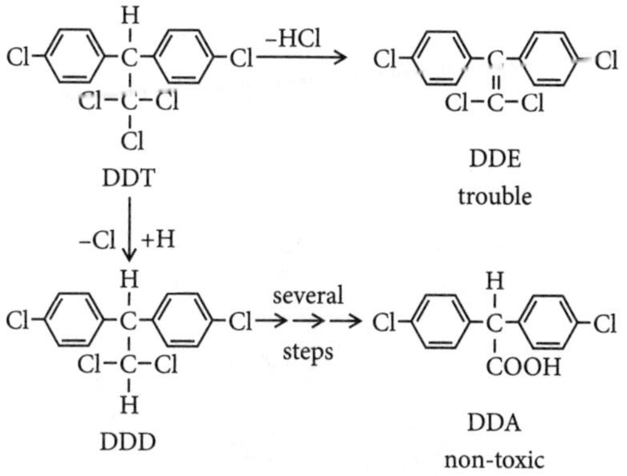

Figure 4.5 Chemical formulas of several DDT compounds. Drawn by Marie Gladwish and reproduced with her permission.

DDT, THIN EGGSHELLS, AND REDUCED REPRODUCTION IN BIRDS

The reproduction of predatory birds became a major issue in the presentations in Madison, overshadowing the direct mortality of smaller songbirds. The latter has a local impact, but reproductive impairment has a far greater, population-wide effect.

Falconry was the sport or art of kings going back for 2,000 years, with records of nesting falcons kept for hundreds of years, forming a baseline for post–World War II populations. Joe Hickey described how falconers from North America and Europe had become concerned about the precipitous decline of their favorite species, the Peregrine Falcon (Fig. 4.6). Peregrine populations had declined severely on both continents since 1950, worse in some regions, less severe in others, unchanged in regions without DDT exposure (e.g., the Aleutian Islands and Scottish Highlands). East of the Rockies in the United States, the bird had become extinct as a breeding species. The problem was reproductive failure, and in every case DDT was present in the eggs and any dead chicks. Other birds of prey, including the Bald Eagle and Osprey, had suffered similar declines under similar circumstances.

© WILL JAMES SOOTER

Figure 4.6 Peregrine Falcons sometimes transfer prey from one bird to the other in midair. With permission of the photographer, Will Sooter.

The serious reproductive problems of predatory birds, especially the peregrine, had become widely known as a result of an international conference held at UW in 1965 (Hickey, 1969). The conference resulted in a 596-page book, *Peregrine Falcon Populations: Their Biology and Decline*, documenting the widespread and severe decline of predatory birds. The conference focused on peregrines, but declines in the Bald Eagle, Osprey, Cooper's Hawk, Sharp-shinned Hawk, Northern Harrier, and White-tailed Eagle also were reported. "Pesticides" were suspected as a cause, but proof did not then exist. Hickey said that reproductive failure of these birds involved egg breakage in the nest, suggesting disturbed calcium metabolism. Soon after the conference, Derek Ratcliffe in England began determining thicknesses of peregrine eggshells in museum collections around the British Isles going back to 1890 (egg collecting by ornithologists had been an important activity half a century earlier).

Even before publication, Ratcliffe reported what he was finding to Hickey, whose graduate student, Daniel W. Anderson, set about determining peregrine eggshell thicknesses in North American museum collections. Their findings startled the ornithological world: Peregrine eggshell thicknesses were stable from 1900 until 1947, when they suddenly became

18% thinner, and remained thinner, in both Europe and North America (Hickey & Anderson, 1968; Ratcliffe, 1967) (Fig. 4.7). Even more startling was the coincidence that DDT had been introduced on a large scale into the world environment shortly after World War II, in 1946–1947. DDT use was immediately followed by eggshell thinning on two continents.

Another coincidence comes from Charles Broley, a retired banker from Winnipeg, Manitoba, Canada, who each year from 1939 banded Bald Eagle chicks in southwest Florida. By 1946 he was banding 150 chicks each year. But in 1947, the same year I first went to Florida and saw my

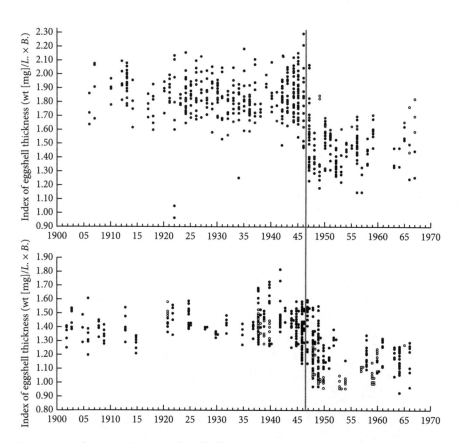

Figure 4.7 Change in the ratio of eggshell weight to size, an index of thickness, in eggshells of Peregrine Falcons (above) and Eurasian Sparrowhawks (below) in Britain, 1902 to 1967. Eggshells became markedly thinner in both species beginning in 1947; learn why from the text. Reprinted by permission from Macmillan Publishers Ltd: DA Ratcliffe, *Nature* 215, 208–210, copyright 1967.

first Bald Eagles, reproduction declined and 41% of occupied nests produced no chicks. In following years 48%, then 60%, and then 78% of nests in 1950 had no chicks. Nesting failure by 1957 was 86%. Broley concluded that the Bald Eagle was becoming sterile (Broley, 1958): "Our American bald eagle—national emblem of this country—is a very sick bird."

The Bald Eagle is probably the best-known bird in North America (Fig. 4.8). Its range is restricted to North America. There may have been half a million of them before European settlers arrived, and perhaps 100,000 breeding pairs in what is now the lower 48 states. Although Benjamin Franklin described it as "a bird of bad moral character," on June 20, 1782, the Bald Eagle became the national bird of the United States, and it has been on the Great Seal of the United States of America ever since. Look at the dollar bill in your pocket, on the podium when the President speaks, on the seals of most federal agencies, and on the lapels of security inspectors at airports (Fig. 4.9). The Bald Eagle has been on American paper money and coins for more than 200 years. The bird fared poorly because of shooting, loss of habitat, and other factors, so in 1940 the Bald Eagle Protection Act became law, protecting both the Bald and Golden Eagles because the immature birds of both species are similar and not

Figure 4.8 Papa Bald Eagle departing on his way to catch a meal for his family. Thanks to Mike Hamilton, photographer.

Figure 4.9 The Bald Eagle is the symbol of America. Official seals of the United States of America (upper right), the President of the United States (upper left), and the U.S. Department of Veterans Affairs (lower left). Lower right: American currency has featured Bald Eagles for more than 200 years. Shown at upper left is a 50-cent coin (1829); the small coin at lower right is a 2.5-dollar gold piece (1928).

immediately identifiable by inexperienced people: The Bald Eagle does not have a white head and tail until it is five years old.

Although protected by federal law, its population continued to decline, and in 1967 the Bald Eagle was placed on the Endangered Species List. A new problem had appeared for the bird. In his testimony Hickey reported that Bald Eagles in Florida exhibited eggshell thinning of 18% to 20%, and their populations were in severe decline because of the breaking of thin-shelled eggs in the nest. Hickey introduced into evidence a photo of a Bald Eagle nest with one live chick and one thin-shelled, broken egg with the shell flaked away from the membrane (Fig. 4.10). That photo ended up on the cover of *Science* on February 7, 1969. Our national symbol

Figure 4.10 Bald Eagle nest with one chick and one thin-
shelled broken egg, entered into evidence in Madison DDT
hearings, December 1968. This photo appeared on the
cover of Science, February 7, 1969.

had become very scarce; existence in the continental United States was
threatened.

Ospreys in New Jersey showed shell thinning of 25%; in Massachusetts
and New Jersey peregrine eggshells had thinned by 21% to 26%. All of this
eggshell thinning occurred after 1947, all of these populations were declining sharply, and the peregrine had been extirpated in the eastern United
States (Hickey & Anderson, 1968). By contrast, the Red-tailed Hawk,
Golden Eagle, and Great Horned Owl exhibited no shell thinning and no
population declines. These species feed on herbivores and therefore had
shorter food chains and less biological concentration of DDT residues.

The coincidence of widespread DDT introduction after World War II
and the onset of eggshell thinning in 1947 was startling to say the least, but
coincidence does not prove cause and effect; only controlled experiments

could prove that. Dr. Hickey also pointed out that as DDE (the DDT metabolite) increased in Herring Gull eggs, shell thickness decreased. He knew he could not prove it (yet), but Joe claimed that "DDE is the compound of extinction."

At that time many of us believed that enzyme induction in the liver of the birds was the cause of the thin-shelled eggs. It was thought to work as follows. The liver in birds and mammals is a detoxifying organ. When a toxic material is encountered, the liver synthesizes enzymes (proteins) to break down the toxicant—that is, to metabolize it. To do so, the enzymes must have a broad spectrum of action to deal with the unpredictable variety of toxic insults that might be encountered. In this case, DDT induces the formation of detoxifying enzymes that also modify the birds' own sex hormones, including estrogen.

In female birds, before egg laying, estrogen secreted by the ovary causes the bird to store calcium carbonate in the hollow parts of her skeleton, especially the femur. At egg formation time, the calcium enters the bloodstream and passes through the oviduct membrane and into the terminal portion of the oviduct, or shell gland, where it becomes the eggshell. Since DDT was correlated with disrupted calcium metabolism in birds, EDF presented witnesses to testify about enzyme induction by DDT.

Enzyme induction in the liver by chlorinated hydrocarbons, including DDT, was for years primarily of interest to biochemists, but recently environmental scientists had discovered that the phenomenon may be ecologically important. Dr. Richard M. Welch, biochemical pharmacologist from Burroughs Wellcome and Company, told how DDT induces the synthesis of enzymes in the liver in a variety of test animals (Conney, 1967). These broad-spectrum liver enzymes can degrade the steroid sex hormones testosterone, progesterone, and estrogen when DDT is administered to test animals at concentrations common in the environment. Welch told how he and colleagues also discovered that DDT can itself function as an active estrogen (Welch et al., 1969), yet another mechanism of action for the material. The estrogenic action of DDT, however, does not replace the bird's own natural estrogen and its function in calcium metabolism.

A Madison *Capital Times* reporter, Whitney Gould, covered the hearings every day and sometimes had dinner with us. Other news outlets around the country often picked up her stories, leading to much national coverage of the hearings. The day after Welch's testimony the headline in the *Capital Times* read "Scientist Warns of DDT Peril to Sex Life." One never knows how the media will handle a story, but that headline certainly got some attention.

It took the testimony of several of us to clarify the environmental importance of enzyme induction. Bob Risebrough and I had pointed out that estrogens affect calcium metabolism and eggshell formation in birds, and that elevated estrogen metabolism caused by DDT-induced enzymes might depress estrogen levels and result in birds laying eggs with thin shells (Hickey, 1969; Peakall, 1967; Wurster, 1969c).

The next witness was Dr. Lucille F. Stickel from the Patuxent Wildlife Research Center in Laurel, Maryland (its director from 1973 to 1982), Bureau of Sport Fisheries and Wildlife, U.S. Department of the Interior. We had known that important experiments were being performed at Patuxent, but we had no idea of the results. We were tense and anxious about what Dr. Stickel would report, but she refused to tell us anything until she was actually on the witness stand. The experiments had been completed only days before her testimony. Our case was developing and changing even as we were presenting it, a repeating pattern in our 10-year effort.

Stickel reported the controlled experiments that established the cause-and-effect relationship between DDT and/or dieldrin and thin-shelled eggs. American Kestrels, falcons, and close relatives of the peregrine were fed environmental concentrations of DDT (2 ppm) and dieldrin (0.33 ppm). Both environmental contaminants were combined in the experiment because they did not have enough birds to separate the variables and still have statistical significance. Eggshells 15% to 17% thinner than normal were the result (Porter & Wiemeyer, 1969).

To separate the variables and discover which chemical caused thin eggshells, the experiment was repeated with Mallard ducks. The ducks had different dietary physiologies but were found to respond to DDE similarly to raptors and were far easier to handle as laboratory animals. When the Mallards were fed 3 ppm of DDE, eggshells 13.5% thinner than controls resulted. (See Fig. 4.7 for formulas of DDT and DDE.) They cracked and broke six times as often and produced half as many ducklings (Heath, Spann, & Kreitzer, 1969). DDT behaved similarly, but DDD did not cause thin eggshells. DDE alone was then shown to cause eggshell thinning in kestrels (Wiemeyer & Porter, 1970).

The direct cause-and-effect relationship between DDT and its metabolite DDE had, finally, firmly been established. *Stickel's testimony was brief but crucial, proving conclusively that DDT and DDE caused thin-shelled eggs and reproductive failure.* DDE was the active shell-thinning agent; DDT caused shell thinning by first being converted to DDE.

The Task Force for DDT made only one attempt to rebut this evidence. Francis Cherms of the UW Department of Poultry Science fed 200 ppm of DDT to Japanese Quail with no effect on eggshell thickness or reproduction rate. Under cross-examination Cherms agreed that he was an expert on herbivorous chickens and quail, but not on carnivorous birds of prey. Both Stickel and Robert Rudd emphasized that chickens, pheasants, and quail were physiologically different from carnivores at the top of food chains (Rudd, 1964).

A reporter for *Science* magazine, Luther J. Carter, had been covering the Madison hearings extensively; he often had dinner with us and attended some of our strategy sessions. With assistance from Carter, the photo of a Bald Eagle nest with a single chick and one unhatched, thin-shelled egg that Hickey had introduced into evidence appeared on the cover of *Science* on February 7, 1969 (see Fig. 4.10), along with an extensive article by Carter in the magazine. Our case was getting more and more media attention. Public education was an important part of our strategy.

DDT REDUCES REPRODUCTION IN TROUT AND SALMON

Evidence that DDT reduces reproduction in fish was presented by Kenneth J. Macek of the Fish-Pesticide Research Laboratory (Columbia,

Missouri), U.S. Department of the Interior. Macek described experiments in which small amounts of DDT, comparable to levels found in forage fish in nature, when added to the daily diet of sexually mature brook trout for more than five months did not kill the adult fish but increased the mortality of offspring (sac-fry) from the DDT residues stored in their egg yolk (Macek, 1968a). A major portion of the mortality of fry in groups where the eggs came from fish fed DDT occurred during the 15th week of development, coinciding with the period of maximum utilization of yolk fat reported to occur in other salmonid fry. He said that DDT concentrations in his laboratory fish were comparable to those from several freshwater lakes, including the Coho salmon from Lake Michigan, where abnormal fry mortality had been occurring.

Risebrough gave similar analyses for fish from the Pacific Ocean, suggesting that important marine fisheries were threatened. Macek also found increased susceptibility to stress (i.e., starvation or reduced food intake) among fish previously fed diets containing levels of DDT comparable to those existing in natural food chains. He reported that mortality during the stress phase of the experiment was 88% among those fish previously fed diets containing DDT, compared to mortality of only 1.2% among control fish (Macek, 1968b). These data and assertions were not contested by the Task Force for DDT.

Meanwhile, during the late stages of the hearings in Madison, developments in Michigan continued to confirm our assertions, even though we were doing nothing further in that state. Coho salmon fry were dying from DDT poisoning in the fish hatcheries, as we had warned in our complaint. In April 1968 Michigan State University had withdrawn its recommendation for using DDT on elm trees, encouraging the cities we had sued to accept court orders that they would no longer use DDT.

On April 17, 1969, Michigan became the first state to ban all sales of DDT. Our legally unsuccessful suit in late 1967 had set in motion a chain reaction, and the dominoes continued to fall. *We had won while losing in Michigan, just as had happened on Long Island.* Taking scientifically well-supported environmental cases to court appeared to be an excellent mechanism for highlighting problems, educating the public, and even getting action on the problem, despite the legal technicalities that were locking the courthouse doors. We dreamed that maybe, someday, we would win by winning in court, but that was destined to be a few years in the future.

WAYLAND HAYES: "DDT IS ABSOLUTELY SAFE"

The original petition for the Wisconsin hearings made no mention of a relationship between DDT and human health, but the subject inevitably became the center of much testimony. Billed as the world's most eminent toxicologist, Dr. Wayland J. Hayes from Vanderbilt University said that DDT was "absolutely safe" for the human population at current exposure levels, based on his studies of men exposed to high levels of DDT who showed "no clinical symptoms" (Hayes, Durham, & Cueto, 1956; Laws, Curley, & Biros, 1967).

Hayes said that 35 men working in the Montrose chemical plant for up to five years making DDT showed no abnormalities in blood and urine analyses, chest x-rays, cancers, and general physical examinations. He did similar examinations on 51 prison volunteers fed DDT for 18 months, again with no apparent ill effects. Storage of DDT and DDE in various tissues was extensively recorded in the latter study.

The testimony of several scientists did not agree with Hayes's conclusions. Dr. Theodore L. Goodfriend of the UW School of Medicine said that "one cannot conclude that DDT is absolutely safe for human use." Hayes's studies were statistically insignificant, they involved only men, and the time frame was much too short to identify numerous potential chronic problems, such as cancer or mutations. He described a variety of possible hormonal effects that have not been investigated and suggested they should be, since DDT is known to interfere with endocrine systems. Earlier, Welch had indicated that DDT at concentrations currently found in human fat was associated with elevated levels of enzymes that break down steroid hormones in rats, and "if one can extrapolate data from animals to man, then one would say that a change in these liver enzymes probably does occur in man." He said that a controlled experiment with DDT had not been done in humans, but an uncontrolled experiment "should not be done on the worldwide population." Hayes admitted he had done none of these tests.

A TOXICOLOGIST ARRIVES FROM SWEDEN

At about this time, in early May 1969, we learned that on March 27 Sweden had announced a moratorium on the use of several chlorinated hydrocarbon insecticides, including DDT, following an extensive literature research by a team of scientists—the Working Group on Environmental Toxicology, Ecological Research Committee of the Swedish Natural Science Research Council. The chairman of the scientists' team was Dr. S. Goran Lofroth of the Royal University of Stockholm. Vic Yannacone issued one of his terse orders: "Get him!" I soon found myself on the telephone to Stockholm inviting Dr. Lofroth to come to Madison to testify. He and I had never heard of each other, I could offer him only airfare, but astonishingly, he accepted. In two days he was in Madison. Robert (Bob) B. McConnell, Public Intervenor and Assistant Attorney General for Wisconsin, had intervened on our behalf; he sponsored Lofroth and paid the airfare.

Interrupting the defense case by the Task Force for DDT (Vic's intentional strategy), Lofroth discussed the ubiquitous worldwide distribution of DDT residues in human tissues and mentioned especially its presence in human milk (Curley & Kimbrough, 1969). He indicated that women excrete a higher proportion of ingested DDT into their milk than do cows and that nursing infants receive about twice the maximum daily intake of DDT compounds recommended by the World Health Organization. Lofroth stated that similar concentrations of DDT caused biochemical changes in laboratory animals.

Lofroth agreed with other scientists that the safety of DDT had not been demonstrated and that its use in the environment should therefore be discontinued. He also quoted publications suggesting that DDT interferes with normal body biochemistry, that it causes tumors in mice

(Tarjan & Kemeny, 1969), and that there is a correlation between higher-than-average residues of DDT and the frequency of human deaths from various disorders, including liver cancer (Radomski et al., 1968). From Lofroth's testimony we gained the impression that European countries had recognized the scientific case against DDT and were taking measures to control it, which had not happened in the United States.

THE TASK FORCE FOR DDT QUESTIONS THE ANALYSES; CONFUSION WITH PCBs

As an important part of its defense, the Task Force for DDT tried to demonstrate that analyses by environmental scientists were in error because there was interference from polychlorinated biphenyls (PCBs), industrial compounds that are also widespread in the environment. To cast doubt on the analyses of several scientists who had testified on behalf of EDF, the Task Force for DDT presented Francis B. Coon, head of the Chemistry Department of the Wisconsin Alumni Research Foundation (WARF), where many of these analyses had been performed. He said that there was interference with DDT and DDD and that he could not be certain there was no interference with DDE.

To clarify, gas chromatography involves injection of a sample into a column packed with a material that causes components of the sample to flow through the column at differing rates. A flow of gas carries the sample through the column. A detector at the end of the column draws peaks as the different chemical components emerge from the column. The Task Force for DDT alleged that the DDT peaks overlap the PCB peaks, confounding the result.

Since Hickey had presented only DDE analyses in his testimony because of the interference problem at WARF with DDT and DDD, the status of DDE became the object of much testimony. During a highly technical, day-long cross-examination by Yannacone, Coon changed several positions he had taken on direct examination, finally admitting that there was no significant PCB interference with the DDE analyses.

Further testimony on analytical techniques and PCBs was presented by Paul Porter of the Shell Development Company, who said that DDT, DDD, and DDE can be distinguished in the presence of PCBs using gas chromatography. Porter also expanded on earlier testimony about the physical and chemical properties of DDT, its degradation, and its transport mechanisms. EDF considered Porter's testimony to be competent, accurate, and in no conflict with our positions; therefore, he was not cross-examined.

In its rebuttal case, the Task Force for DDT had replaced McLean as its attorney with Willard S. Stafford, a well-known Madison attorney. Stafford then called Bob Risebrough as an adverse witness for additional questioning about his recent paper "Polychlorinated Biphenyls in the Global Ecosystem." It had suggested that PCBs might have a role in the thin-eggshell phenomenon. Bob had been sitting next to Vic Yannacone suggesting questions to be asked and was therefore immediately accessible to be called to the witness stand. He had attended almost all sessions of the hearing and McLean had cross-examined him for three days in December.

Stafford was obviously trying to get Dr. Risebrough to say that it was PCBs, not DDE, that were causing eggshell thinning, but Bob was ready for that one. While in Madison, Bob was the guest of Dan Anderson, Hickey's PhD student, and they had studied Dan's original chromatograms and calculated correlations among PCBs, DDE, and shell thinning in Double-crested Cormorants and White Pelicans. The chromatograms were made in connection with Anderson's PhD thesis. They concluded that although PCBs were powerful enzyme inducers in the birds, they did not correlate with shell thinning of White Pelican eggs. That was the first time PCBs had been shown not to cause shell thinning, a conclusion confirmed repeatedly thereafter.

Bob almost by chance had a manuscript of this new paper in his briefcase, which he then entered into evidence (Anderson et al., 1969). He concluded that it was DDE, and not PCBs, that was largely responsible for the widespread occurrence of thin eggshells among carnivorous birds. That was not the evidence Stafford was seeking. Bob also used the opportunity to present new evidence on the almost complete reproductive failure early in 1969 of the Brown Pelicans on Anacapa Island in southern California, whose very-thin-shelled eggs collapsed when the birds tried to incubate them, leaving the nesting colony littered with broken eggshells and almost no chicks (Risebrough, Sibley, & Kirven, 1971).

Appearing on behalf of the U.S. Department of Agriculture (USDA) was Harry W. Hays, director of the Pesticides Regulation Division. He described the registration process: USDA has complete authority for registration while other agencies have only advisory capacity. Chemical companies apply for the registration, but USDA does not check the accuracy of the data. He said there had been new registrations for DDT, but data on sublethal effects on animals or mobility of DDT had not been required for registration. Under the law, only the Secretary of Agriculture could then initiate pesticide cancellation proceedings.

Another DDT proponent was Samuel Rotrosen, the president of the Montrose Chemical Corporation, the largest DDT maker in the United States. The Montrose plant in Los Angeles made about half the U.S. total. Other companies making DDT at that time were Diamond-Shamrock, Olin Mathieson Chemical, Allied Chemical, and Lebanon. Of 114 million pounds made in 1968, about two thirds was exported. In large lots DDT cost at that time about 17 cents per pound; its total value in 1968 in the United States was $20 million.

INTEGRATED PEST MANAGEMENT WITHOUT ANY DDT

Testifying for the Task Force for DDT were two entomologists, R. Keith Chapman of the UW Department of Entomology and Bailey B. Pepper of the Department of Entomology of Rutgers University. Chapman said that DDT was still needed to prevent insect damage to cabbage and carrots in Wisconsin. Pepper said that DDT was needed to prevent damage to several crops and to prevent mosquito-borne encephalitis, but he agreed that malathion (an organophosphate insecticide) was a good substitute and that DDT was not recommended for use on marshes.

In emphasizing the alternatives to DDT, EDF presented three top scientists in the sophisticated field of integrated pest control, the blending of biological and chemical techniques into an integrated system. Dr. Robert van den Bosch, entomologist in the Division of Biological Control, University of California at Berkeley, described ways in which an agro-ecosystem can be manipulated "to manage pest populations so that they do not cause economic loss" (Smith & van den Bosch, 1967). He discussed the role of entomophagous insects (insects that are parasites and predators of other insects) in controlling potential pest species and pointed out that DDT often causes a rapid rebound of the pest populations after eliminating the slower-rebounding entomophagous insects.

Dr. van den Bosch told how for many years he had recommended DDT for some purposes, but that more recent knowledge of its enormous ecological impact had caused him to discontinue these recommendations. He described DDT as an "ecologically crude material . . . developed by chemists and toxicologists . . . with no ecological thought whatsoever" and "exploited largely by people who were thinking in terms of their economics. . . . Most entomologists eagerly seized" the material "totally ignorant of the . . . ecological implications" of its use. Of those implications, he said, "I'm scared." EDF had established in these proceedings that DDT is toxic to nontarget insects and animals, persistent, mobile, and transferable, and that it builds up in the food chain. No label or directions for use could change these properties.

Testifying on May 21, Dr. Paul DeBach, eminent insect control specialist of the Department of Biological Control, University of California at Riverside, and Donald A. Chant, chairman of the Zoology Department at the University of Toronto, reiterated van den Bosch's positions. DeBach (1964) described DDT as a highly disruptive material in an agro-ecosystem and told how it causes outbreaks of mites and scale insects by killing their natural enemies. Chant (1966) discussed the concept of economic threshold, pointing out that insecticides are often used when the pest population is below a level of economic damage, or even totally absent from the area. "DDT," he said, "has no place in integrated control."

All three of these scientists, Drs. van den Bosch, DeBach, and Chant, were world-class scientists in integrated pest management, and all three came to Madison as volunteers to testify in the hearing.

SUMMATION: DDT IS A POLLUTANT IN THE WISCONSIN ECOSYSTEM

The DDT hearing in Madison was remarkable for the great diversity of disciplines that played a role in its organization and presentation. That was especially true for the summation, which was preceded by a week of intensive work and dialogue among mathematicians, various ecologists, systems analysts, and engineers, all with the searching probes of Vic Yannacone. The result was a brilliant summary of available data on DDT transport, uptake, and metabolism in the form of a complex, modern systems analysis that left the Task Force for DDT incapable of critical cross-examination.

Summation testimony was given by Drs. Orie Loucks and Robert Rudd. Following his systems analysis presentation, Dr. Loucks concluded that DDT concentrations at higher trophic levels in the Wisconsin ecosystem could be expected to increase, with further decreases in numbers of important predator species, leading to instability and degradation of the ecosystem.

Rudd (1964) summed up the problem by saying that pest control operations too often have been restricted to a consideration only of a particular pest and a particular crop. "The pest control operator, once the applications have been made, pretty well forgets the problem. . . . The ecologist, on the other hand, is concerned with entire . . . ecosystems. He has no particular restrictions." Both Rudd and Loucks said that DDT fits the definition of a pollutant in "Wisconsin law since it is deleterious to fish, bird, animal or plant life."

This interdisciplinary systems analysis and model became a major paper published in *Science* a year later (Harrison et al., 1970). Since then this modeling technique has been expanded to make projections of the distribution, uptake, and metabolism of many different pollutants, metals, and radioactive isotopes in the United States and around the world. It has become a widely accepted mechanism for forecasting the long-term outcomes within regional ecosystems of persistent chemicals and radioactive isotopes and their impacts on organisms at various trophic levels (Loucks & Leavitt, 1999). This great advance began with the Madison DDT hearing.

TO MAKE AN OMELET, YOU WILL BREAK SOME EGGS

We made numerous friends during those hearings in Madison, but less visible were those who didn't think that what we were trying to do was such a good idea. *Seeking to ban or reduce sales of a profitable product is not a good way to make friends with its maker.* We were seeking bans on DDT and several other chlorinated hydrocarbon pesticides, which stirred up the animosity of the entire pesticide industry. Emphasizing that we were not after all pesticides, only this destructive handful, did not seem to matter.

The May 5, 1969, issue of *Barron's* ran a two-page article with the title "Up with People, and Down with the Venomous Foes of Chemical Pesticides," which summarized the DDT proponents' (losing) case as presented in the Wisconsin hearing and as would come later in Washington, DC. It claimed that DDT helped eliminate malaria from the United States, which it did not; malaria was essentially gone from the United States before DDT ever arrived. The article threatened famine and disease as "a grim portent of what might happen if the know-nothings ever get their way" and "If we had to depend on nature . . . we would probably die of disease at a fairly early age, if we did not starve first." We (EDF and environmental scientists) were allegedly attacking all of agriculture and endangering the food supply, even though the major uses of DDT at that time in the United States were on elms and cotton (cotton represented 86% of the total).

Farm Chemicals (January 1969) said we were a "parade of beards" (several of us did have beards), "a strange assortment of characters," and an "arrogant collection of lawyers and pseudo-scientists . . . [who] would abolish a great system and offer nothing in return [except] chaos." *Farm Technology* (January 1969) said that "EDF knows that in a scientific hearing, their so-called scientists would be laughed out of the room. That's why they resort to legal hocus pocus!"

The July 6, 1970, issue of *Barron's* had another two-page article entitled "Ravaged Summer, It's the Natural Sequel to 'Silent Spring,'" apparently designed to scare the wits out of the public. Here are some excerpts:

"If they [EDF] succeed, their triumph will be shared not only by the gypsy moth, but also by the rednecked cane borer, climbing cutworm, carrot weevil, cabbage looper, onion maggot, darkling beetle, white grub and the rest of the estimated 210 insect pests for which, in most cases, DDT is the sole known means of control. The nation's farm and wood lands and food supply, not to mention health and welfare, contrariwise, might not recover. Win or lose on the issue, the nature-lovers already have left their mark on the landscape. In less than a decade, 'Silent Spring' has spawned ravaged summer."

"*Barron's* time and again has sought to debunk the extravagant charges and wild alarmism over DDT, which have gained spurious circulation in a 'largely rigged market for ideas, where anti-capitalism is the rage and anything that smacks of it, no matter how outrageous or absurd, can count on an incredible longevity.'"

"What the environmentalists are seeking would be a major disaster, both at home and abroad. To deny the product to such nations as India would constitute, in the words of one scientist, 'an act of genocide.'"

"There is no satisfactory substitute for DDT in the control of rattlesnakes in the Southwest U.S."

"Through their unbridled recklessness with facts and sheer irrationality—a triumph of superstition over science, . . . they are threatening to unleash famine and pestilence upon their fellow citizens. They profess to preserve wildlife, defend the environment, befriend the earth. Their natural prey is civilized man."

We had not heard that bit about rattlesnakes before, nor did we again (we thought DDT was an insecticide, not a "snakeicide"), but the "anti-capitalism" and "genocide" charges would return—decades later!

We were compared with Charlie Brown's Linus: "I love humanity; it's people I can't stand."

Character assassination is part of the game. The following passage was from Sponsors of Science—DDT, circa 1970:

> Dr. Wurster . . . one of the master minds behind the disgraceful Wisconsin DDT Hearings, . . . is well-known within the scientific community for his vitriolic attacks on biologists who disagree with him. . . . [Wurster was] instrumental in founding the now extremely powerful and wealthy Environmental Defense Fund.

Bob Smolker's "Mystique Committee" apparently was succeeding in its mission of puffery, convincing our opposition that we were vastly more than reality at a time when we didn't even have a letterhead.

Anonymous and undocumented chemical industry representatives "cast Wurster as a radical ecologist out to destroy the American way of life. Wurster's politics seemed, to a great many Americans, downright dangerous" (Kinkela, 2011). It was about 40 years ago that a totally false and fabricated story was planted in which I allegedly made highly inflammatory and racist remarks at a news conference. This outrageous story found its way into the *Congressional Record* and has been scattered about the Internet ever since. There was no such news conference or remarks, but denial makes no difference and the story will live forever. Character assassination works that way.

On the one hand, these attacks against us were annoying, but on the other hand, they suggested that we were making headway against the DDT problem. Being ignored would have been worse.

The DDT hearings in Madison, Wisconsin, began on December 2, 1968, ended on May 21, 1969, nearly six months after they began, included three recesses, involved the testimony of 32 scientists and others, consumed 27 days, and generated a transcript of 4,499 pages and 208 exhibits weighing 40 pounds. It was a tremendous, truly grassroots effort by an extremely dedicated team involving hundreds of purely volunteer scientists and other citizens, all aiming toward the ultimate goal of not only a state but also a national ban on the insecticide DDT.

EDF came away from it with the sense that we were going to win in Wisconsin—that is, that DDT would be declared a pollutant of the state's waters. We had presented a full and comprehensive case by expert witnesses, the ones who had actually done the work, and the industry's rebuttal was weak and unconvincing, at least to us. The hearing examiner, Maurice Van Susteren, had been comprehending of the science and fair in his rulings. We believed that he would rule in our favor.

MID-1969: WHERE DID WE STAND, WHAT HAD WE ACCOMPLISHED?

So what was accomplished with all this effort, and what still needed to be done to reach the goals? First and foremost was a very substantial amount of public education from our several court cases, and especially from the hearings in Madison. National media coverage ranged from major newspapers—*The New York Times, The Washington Post, Los Angeles Times, Chicago Sun-Times,* and *The Wall Street Journal*—to many national magazines, including *Reader's Digest, Saturday Review,* and *Science,* most of them accurately reflecting the scientific issues we were describing. National television coverage was also extensive.

The general tone of this publicity was that we were a group of well-prepared and competent scientists versus a poorly prepared industry with weak excuses for their product. This quote from *The Wall Street Journal* of March 4, 1969, was typical:

The industry is well organized to defend itself, but the Wisconsin attack seemingly caught it off balance. The task force didn't retain an attorney to represent it at the hearings until less than a week before they opened, and it apparently had almost no idea of what it would face in Madison. "Frankly, nobody knew what kind of hearing this was," says Louis A. McLean, the attorney finally picked by the task force. "We thought it would be something like a legislative hearing, where people get up and make statements of position." Mr. McLean is a long-time industry spokesman, who until he retired in July, 1967, was secretary and general counsel for Velsicol Chemical Co., a pesticide maker.

In the past, pesticide manufacturers sometimes have tried to dismiss their critics as food faddists and neurotics. For example, in 1967 Mr. McLean wrote: " . . . the antipesticide people in almost every instance hold numerous beliefs in nutritional quackery and medical quackery, and they oppose public health programs." The characterization rankles many of the scientists involved here [in Madison], leading some to suggest that the industry this time has underestimated its opposition. Indeed, [EDF] spearheaded the case against DDT with expert testimony from reputable scientists brought in from all over the country.

This illustrates the importance of our insistence on judicial rules of evidence where witnesses had to be qualified as experts before testifying, with their statements then subject to cross-examination by opposing attorneys. *We didn't want lobbyists and unqualified people making bald, unsupported, and inaccurate statements, then walking away, leaving a confusing, incompetent, and unchallenged record that would produce an unsatisfactory outcome.* Vic Yannacone demonstrated a substantial grasp of the science behind our case, and he was brutally effective at nullifying incorrect or deceptive statements by opposing witnesses. The result was a clean and accurate record of the hearing, along with widespread public education on the topics at issue.

FINAL VERDICT FROM THE DDT HEARINGS IN WISCONSIN

One year after they ended, on May 21, 1970, the hearing examiner, Maurice Van Susteren of the Wisconsin Department of Natural Resources, rendered his ruling on the comprehensive hearing on DDT held in Madison. Van Susteren ruled that "DDT and its analogs are . . . environmental pollutants within the definition of . . . Wisconsin statutes, . . . contaminating and rendering unclean and impure the air, land and waters of the state and making the same injurious to public health and deleterious to fish, bird and animal life." He ruled that "no concentrations, levels, tolerances, or amounts can be established as safe" and that any amount of DDT and its metabolites in the environment is potentially harmful to humans and of public health significance. He labeled DDT an uncontrollable material. Van Susteren's decision, in effect, set a zero-tolerance level in the environment as the only safe level for the protection of human health and wildlife resources.

EDF had conducted the petitioner's case against DDT and had brought scientists from across the country (and abroad) to give testimony. The decision followed an examination of the DDT issue that was probably the most exhaustive ever held for a pesticide. The information generated by the hearings ultimately resulted in restrictions on the use of DDT in Wisconsin and many other states.

In subsequent actions, EDF submitted this ruling to the USDA, to the U.S. Department of Health, Education and Welfare, and to the pending EDF litigation on DDT in the U.S. Court of Appeals for the District of Columbia in Washington, DC, to supplement the already extensive documentation it had provided.

5

EDF, Barely an Organization, Getting Its Act Together

The late 1960s and early 1970s was a world of increasing political unrest on many fronts. In January 1969, Richard Nixon replaced Lyndon Johnson as president. Public support for the war in Vietnam was diminishing and there were widespread antiwar demonstrations.

Environmental awareness and concerns were rapidly increasing. Air and water pollution were increasingly severe. A huge oil spill dumped 100,000 barrels of crude oil onto the beaches of Santa Barbara, California. The Cuyahoga River in Ohio caught fire. Students buried automobiles on college campuses. Lake Erie could no longer support fish. The great whales were being killed in record numbers. People were apprehensive about pesticides. The Bald Eagle, national symbol, was disappearing. The first Earth Day was launched in 1970.

Responding to this public outcry, the National Environmental Policy Act passed Congress almost unanimously and became law on January 1, 1970; the Clean Air Act became law in 1970, the Clean Water Act in 1972, and the Endangered Species Act in 1973; and the Federal Insecticide, Fungicide and Rodenticide Act was rewritten in 1972. Rachel Carson's *Silent Spring* had appeared in 1962 and generated a sizable public reaction, but pesticide policies had changed very little by 1970.

This was the milieu in which EDF sought to pursue its goals of a national ban on DDT and the development of environmental law. Reaching those goals would require a much more substantial organization than EDF was in 1969; at that time it was little more than a board of trustees with plenty of ideas but no staff, no office, and almost no money. Most of those trustees were going about their normal lives with EDF concerns more like a hobby than a profession. Their dedication was strong and very real, but a strategic game plan was barely in sight.

There were additional impediments when compared with today's world. Forty-five years ago communications barely resembled what we have now. Most television sets were black-and-white with small screens and large bulky bodies, although color TV was arriving slowly. There were no computers or cell phones. Telephones were attached to wires, so you didn't wander around the neighborhood while talking on the phone. E-mail did not exist. To send a message to a distant person, you hand-wrote or typed a letter using carbon paper if you needed a copy. The letter then went into an envelope with a stamp, and the post office delivered your message a few days later. Airmail required extra postage. If you could not use the telephone, receiving a response could take days to weeks. International correspondence took much longer.

Preparing documents on a typewriter could be frustrating. Letters, words, or paragraphs could not be changed or moved around. Instead you used Wite-Out, an eraser, or scissors, and if that didn't work you retyped the whole thing. Rapid typing with all ten fingers was a great asset, and nobody had or needed smart thumbs.

This was the world of 1969 in which EDF was ill equipped to pursue its goals. Courthouse doors were still locked by the "standing" barrier, presenting a substantial impediment to pursuing the DDT case or any other environmental issue. We had gained considerable momentum and attracted much attention from our actions in New York, Michigan, and

Wisconsin, which in turn led to EDF's involvement in several non-DDT actions (described in the next chapter).

The summer of 1969 saw little action on the DDT front. Inaction by EDF on DDT could soon mean the DDT effort and EDF itself might quietly fade away. We needed new actions to keep the DDT issue and EDF alive or all our gains could be lost. Trustees were scattered, and I was frustrated.

The obvious answer was to take legal action against the USDA in Washington, DC, seeking a national ban on DDT (USDA was in charge of pesticide regulation). We didn't know how to do that, and doing so would require a stronger EDF. That move sounded intimidating to us and apparently also to Vic Yannacone, who had little experience in federal litigation at that time.

We had the scientific case well in hand and we knew well how to present it. George Woodwell had handed over the Scientists Advisory Committee chairmanship to me. The committee numbered more than 200 scientists by then. This was a very loose sort of committee that would never meet, but its members had agreed in writing to supply expertise and testimony without any fees—a strictly volunteer committee. I regularly kept members informed of developments and often solicited their comments and formal submissions to USDA and other agencies.

EDF lacked the organizational structure to undertake such a federal campaign because our attention had consistently been on solving the problems and attempting to win battles, not on building an organization. It would have been more accurate to call EDF a "group," as the media often did. Our board of trustees had shrunk to nine with the resignation of Tony Taormina, but it went back to ten with the addition of Roland C. Clement, the vice president of the National Audubon Society. Roland finally felt safe to join the board by this time, since environmental litigation had gained some respectability. (Roland, in good health and sharp wit, was 102 years young on November 22, 2014.)

During the Madison hearings George Woodwell had been in contact with Gordon Harrison of the Ford Foundation, which was interested in encouraging the development of environmental law. I doubt the Ford

Foundation fully trusted those wild men from Long Island, and we still did not have a tax exemption, but EDF was one of the few groups actually practicing environmental law. Ford solved the problem by making a grant of $100,000 to the National Audubon Society, with instructions to use the money to support EDF. When we needed money, we called Audubon's comptroller, George Porter, and he sent us a check.

Early in 1969 EDF hired Joseph Hassett, a lawyer and former Jesuit priest from Fordham University, to be EDF's first executive director. EDF still had no office, so Hassett worked out of Vic Yannacone's law office in Patchogue on the South Shore of Long Island. Hassett's tenure was short. By late 1969 he had announced his resignation, promising only to try to identify his own replacement. The office was apparently more chaotic than he preferred.

We were burning up our remaining Ford funds and Vic Yannacone had returned primarily to his private personal injury and workman's compensation law practice. In addition, disagreements and frictions had developed between Vic and the board. Vic had little experience in federal litigation and procedures, which we expected would be in our future. Late that summer we had a re-election of the board of trustees to establish terms of office, and Vic and Carol Yannacone were not re-elected. So there we were with only eight trustees, no lawyer, no staff, no office, no tax exemption, and running out of money. The situation seemed desperate, but nobody was paying much attention. Maybe the whole thing would dry up and disappear.

More changes were in the wind. In October 1969 Joe Hassett and Bob Smolker went to Washington, DC, in search of a new executive director. They had to hire a top-quality director to preside over a staff of zero, with a vanishing budget, no fringe benefits, no security, and an unknown job description. It was a tall order, but there was still that good idea (science plus law) plus a handful of trustees who believed in it.

Joe and Bob must have been super-salesmen, for they came back with not one but two outstanding young lawyers as candidates. The developing new field of "public interest law" was apparently attractive, and EDF, such as it was, offered one of the few games in town to work in that field. One of the candidates was Roderick A. Cameron, a West Point graduate with a law degree from the University of California at Berkeley. He had served in the general counsel's office of the Federal Aviation Administration and had just completed a one-year judicial clerkship at the Court of Appeals in Washington, DC. Rod hastened to Long Island to interview with EDF's executive committee.

At that point, EDF's board of eight trustees was run by an executive committee of only four: Dennis Puleston as chairman, Art Cooley, Bob Smolker, and me. There wasn't really that much to "run." The interview with Rod clicked well with all parties, and in the fall of 1969 he became executive director (Fig. 5.1). Rod commanded instant respect because he was so tall that he had to duck to get through the doorways. But it was a tall order for even a tall person to drag EDF out of the dustbin, where the

Figure 5.1 Rod Cameron, EDF's executive director from 1969 to 1974. Photo about 1972 with his permission.

diminishing bank account still resided with the National Audubon Society. We had an excellent scientific reputation as an activist organization, but nobody was sending money.

The other candidate was Edward Lee Rogers, a tax attorney with the Department of Justice. As a sense of stability developed in the early days of 1970, Lee also was hired to be general counsel. Rod and Lee both moved their families from Washington, DC, to the Stony Brook area on Long Island.

ENVIRONMENTAL DEFENSE FUND, OR FUNDLESS ENVIRONMENTAL DEFENDERS?

We have described the DDT wars as being fought through May 1969, but it was unclear whether EDF was actually going to become an organization or merely remain a small band of troublemakers from Long Island. The battle to curtail the use of DDT cannot be separated from the growing pains, failures, and successes in building an organization to fight these battles.

As the new decade began (1970), we had a new executive director, a general counsel, a secretary, and a one-room office (more about that later), but we were consuming what remained of the Ford Foundation grant. Some called us FED, the "Fundless Environmental Defenders," only partially in jest. We had stirred up a considerable storm in Wisconsin and the first Earth Day was nearly upon us, so we were inundated by requests for interviews, speeches, advice, and assistance with numerous environmental issues—but few checks.

In late 1969 Executive Director Rod Cameron ran the following ad in *The New York Times*:

> Executive Secretary for scientists and lawyers in environmental organization. Fast accurate typing a must, shorthand, some bookkeeping and clerical duties. Salary $140/wk. Reply to P.O. Box 740, Stony Brook, NY 11790.

From among several applicants, Marion Lane Rogers (no relation to Lee) could "type like the wind" but did no shorthand, nor did she have bookkeeping experience. Instead she added to her résumé that "the danger that I will elope with that divine staff bachelor, and retire, or request seven months maternity leave at an inopportune moment is remote—and diminishing." Her friends warned her: "You're not going to send that?! You'll never even get a response." But she did.

Rod called her on the phone: "I liked your résumé. Would you like to come in for an interview?" "I liked your ad. I'll be glad to come." And she did. Marion and Rod hit it off immediately and forever after. Marion brought talent, dedication, wit, class, competence in writing, editing, and typing, compassion, and a great deal of common sense to EDF, which needed them all. Marion started work immediately on the first workday of the new decade, 1970. She remained with EDF for 20 years, then became editor and an author of *Acorn Days*, a book about EDF's early evolution. As of 2014, she is 96.

Rod rented a single very large room in the attic above the Stony Brook post office that became EDF "headquarters" (Fig. 5.2). Marion described the new office in *Acorn Days:*

> The central feature of the Village (Stony Brook) is the post office and under its eaves a life-sized replica of an American eagle flaps its wings to count the hours. An appreciative audience usually gathers at noon to applaud its daily premiere performance of twelve flaps. EDF's office was in the eyrie directly behind that majestic eagle. . . . Promptly at noon not only did our pet eagle flap its wings but the earsplitting twelve o'clock siren from the nearby firehouse also sounded off, so shrill and loud it vibrated the very soles of our feet and all action ceased.
>
> The décor of our one-room office could most charitably be described as "shabby Goodwill": three secondhand desks, some chairs, a couple of used file cabinets, a small, totally inadequate duplicating machine, two secondhand IBM typewriters and an old manual one.

Figure 5.2 Where it all started: EDF's first office (1970) was behind the great eagle at the post office in Stony Brook, New York. Photo by author.

And papers. Papers stacked on the desks, papers piled on the floor, papers on top of the file cabinets. Papers everywhere but in the file cabinets. Who had time to file?

Despite the chaos and lack of amenities it was fun working there. I liked and admired Rod and Lee and the trustees immensely. We were all true believers united by a common cause more important than any one of us. Every EDF triumph called for team rejoicing because somehow we had all played a part in its conception, development or execution.

Rod's desk was buried in papers (see Fig. 5.1) and he was heard to remark "This is a very big job." He had presumed it to be a more-or-less 9-to-5 job, but had not anticipated that sometimes it would be 9 a.m. one day until 5 a.m. the next. Rod's initial assignment was to secure a tax exemption, and by early 1970 we had become a Section 501(c)(3) tax-exempt organization. He also had to soothe ruffled feelings between Vic and the board in connection with Vic's separation from EDF.

So here from *Acorn Days* are Rod's early impressions of EDF soon after his arrival, early in 1970. He can tell it better than I:

EDF had one extraordinary asset: its "Founding Four" Executive Committee. These were a high school biology teacher, a wonderful Old World naturalist-adventurer, both from the South Shore of Long Island, and two Stony Brook University professors, one in biology and the other a chemist in marine sciences. All were avid bird watchers. They were passionately concerned about the effects of DDT upon their fine feathered friends and on the ecosystem. They were also passionate in their belief, fashioned partly by their association with Yannacone, that a partnership of scientists and lawyers could accomplish something the eloquence of Rachel Carson could not: regulatory reform that would change the pesticide practices of agribusiness.

Their commitment and their native abilities overcame all their lacks in the usual attributes sought in board members of such an organization: fundraising contacts, government contacts, publicity contacts, managerial skills, and so on. Of wealth, wisdom and work, the standard trustee assets, they contributed the last in abundance. Of wisdom they contributed their substantial knowledge of the scientific community and of the ills they sought to address. Of wealth they had none.

In addition to the work which they contributed so unflaggingly, they communicated an enthusiasm which constituted an ever present force field within which EDF operated. They nurtured EDF from its early fluttering to its later steady, strong flight. In view of the fact that they had jobs and families to support and love, their commitment was astonishing.

Another primary moving force in that original group was Victor Yannacone, a South Shore personal injury and workers' comp lawyer.

He had a flair for publicity and for the dramatic gesture. He was litigious and quick on his feet. Much credit for EDF's early emergence as a champion of a new approach to environmental reform should be given to Yannacone. He was combative, energetic, imaginative and irreverent. He had enough chutzpah that he easily would have brought a class action suit against the Deity.

Yannacone was a quick study on scientific matters. Whether he fully understood all the scientific evidence that was passing through his case, the opposition would rarely know and, even more rarely, have time to do anything about. He was a slugger and if some of his punches went wild, many of them hit home with telling effect. His vaulting ego and abundant energy, harnessed (somewhat) to the discipline of EDF's trustee-scientists, provided the thrust to get EDF airborne at the dawn of what some now refer to as the Environmental Era.

EDF's earliest approach sought to harness careful scientific doctrine behind whatever flaky legal theories it could cook up to get itself a forum and a pulpit. Before 1970, of course, few of the environmental laws and regulations we now have were on the books. Legal causes of action on behalf of the environment had to be derived from the interior reaches of the imagination. EDF complaints of the pre-1970 era showed that origin. Cross examination was supposed to overcome the weaknesses of our legal theories. There was much talk in EDF about the crucible of the adversarial process burning away the evasions, temporizations and data-tampering of the polluters and of the cozy-with-industry regulatory agencies.

Cross examination was a Yannacone strong point. The problem was to survive the opposition's preliminary motions to dismiss . . . and to win a hearing in which the "enemy" could be cross examined. Yannacone probably accomplished more in that pre-1970, pre-NEPA [National Environmental Policy Act] legal environment than many a more careful attorney would have. [He] was the EDF champion-at-arms.

Rod soon told us what we already knew: that we needed more funding or we would not be long for this world. With our newly acquired tax exemption, contributions would be tax-deductible, so we decided to launch a public membership. How does one do that? We didn't know, but a full-page ad in *The New York Times* seemed like a good idea.

That human milk was contaminated with DDT had been known since 1951, and in an article in *Saturday Review* (Wurster, 1970) I pointed out that the concentration was two to four times higher than the tolerance limit permitted by the Food and Drug Administration in cow's milk shipped in interstate commerce. I remarked that *if mother's milk were in any other container, it would be banned from crossing state lines.* My comment got into some newspapers, and I was deluged with letters, not all of them friendly.

One nursing mother from Stony Brook was Joan Ames Woodcock, married to John Woodcock, a graduate student writing his PhD thesis in English at the university. Joan's parents, Amyas Ames and his wife, Evelyn, soon joined as a single member of EDF's board of trustees—two people, one seat on the board; we were always careful to supply them with a very large chair at meetings. It worked well. Amyas was a former board chairman of both Lincoln Center and the New York Philharmonic. With a background in investment banking, the arts, and environmental concerns, Amyas helped launch EDF's financial reporting and bookkeeping systems, about which the rest of us knew nothing.

We decided that a photo of Joan nursing her baby would make a good centerpiece for our *Times* ad announcing our new public membership. Here are some of Joan's reactions from *Acorn Days* as she, John, and I sat around their dining room table writing the ad:

It is dangerous to get between a mother and her baby. My one driving desire . . . was to get the bastards who had been responsible for putting DDT into my body. Little did we know that the battle was already under way and that the good guys all lived down the street. I also have a vivid recollection of the final ad-writing session that [we] survived at our dining room table. [We] sensed that the final draft had to come then, that day, and it had to be perfect. We read the text out loud again and again, and could find not one word out of place. That was a high point in my life.

We gambled about a quarter of our diminishing bank account, then around $25,000, on that ad (Appendix 2), but it brought in more than it cost, unusual for such ads, and it successfully launched our public membership. EDF's finances began to improve from then onward, and nobody was going to lose his or her job. To the contrary, EDF was soon able to hire new staff and expand its programs.

One might wonder just what motivational force was driving these people to do all the things they were doing. It certainly was not money: Most of the team members were volunteers, and the growing staff had paltry salaries and no fringe benefits, and all were well educated and could have earned more money elsewhere. Conservationists of that time always put great faith in education: When the kids grow up, they will fix the world's problems. We did not have that kind of faith, and most important, we did not have that kind of patience.

The adversarial element was unifying. We had found a way to force the "enemy" to read and consider our documents, and then to respond. That was something new to us, relieving the frustration of being ignored and powerless. Many of us were scientists, and we thought we had some environmental solutions that ought to be considered, if not adopted. We didn't think public officials were doing a good job protecting environmental values, and we figured that suing the bastards would make them do better. Furthermore, it was fun (sometimes). Our good team was getting bigger and better, and we were going to save the world—at least part of it.

The social mix helped make the whole operation fly. We had great parties at the Smolkers', with Rosemary Smolker putting out a pyramid of scallops marinated in lime juice while Bob mixed great drinks at his bar. A pig roast at the Pulestons' was always memorable because of their several exotic pets, as well as the food. A Turkey Vulture named Alger (because he hissed) sometimes sat on Dennis's shoulder and preened his hair. Dennis warned that his pet cockatoo had taken off the finger of an earlier guest, so we were always careful to maintain a distance from his cage. Nancy Cooley was a professional caterer, and the Cooley parties were always something to anticipate and remember fondly.

Trustees and staff were a single team, and optimism and resolve were rampant. We were one family, with each member contributing to the whole. We all figured it was better to fight and lose than not to fight at all. We may have been fundless environmental defenders, but we had lots of interesting and pleasant social memories going forward. Furthermore, we were winning some of our goals, even when we were losing in court. Rod Cameron and his team had snatched victory from the jaws of disappearance, and EDF was showing signs of permanence and might soon be called an organization. It was a splendid and enthusiastic team of talented and dedicated people.

6

EDF Diversifies into New Environmental Arenas

Environmental law was essentially nonexistent in 1969, and it was a major goal of EDF to establish, enhance, develop, and use this new strategy for solving environmental problems, not just involving DDT but other issues as well.

It wasn't long after incorporation before EDF was getting much publicity because of its actions, leading to numerous requests for advice and assistance in connection with a variety of environmental problems, along with invitations to become involved in an assortment of issues. Whatever it was or wasn't, EDF certainly was not all talk and no action. There was clearly plenty of action, which attracted plenty of attention. Conservationists were tired of losing by being reasonable, compromising, and timid. Earth Day was about to arrive, and it was time for action. We also were learning that being a "fund" caused a few problems of its own. Some thought it might be a source of funds for them. A few wondered if we were some sort of mutual fund, so Bob Smolker suggested we sell shares in our fund, which would "pay" negative dividends. His Mystique Committee had many original ideas.

For the Long Island trustees and small staff, there was little risk in becoming involved in new cases. We had little money to spend or lose, we had an apparently good idea to pursue for environmental protection, and if somebody sent plane fare, we were on our way. So it was with Clancy Gordon of Missoula, Montana. In fact, he came to us only a few months after EDF's incorporation.

EARLY BATTLES

EDF Takes a Polluting Paper Mill to Court in Montana

It was "Leap Day," February 29, 1968, when about 100 women of Missoula gathered at the gates of the Hoerner-Waldorf pulp mill west of town to protest Missoula's "stinky air." Sometimes the air was so smoggy that planes could not land, and cars turned on their lights in mid-day. "GASP" read one of the picketer's signs, "Gals Against Smoke and Pollution." Other signs said "Phew!," "Bad Sky Country," "Our Air Stinks," "How High is the Big Sky," "Where's the Airport?," and "O, Say Can You See." The demonstration was orderly and peaceful. Hoerner-Waldorf officials got the message and provided coffee and donuts. That was the first salvo in the coming battle to clean up the filthy air of Missoula, site of the University of Montana. "Picketing Won't Help Control Air Pollution," said an editorial in the local newspaper. "We hope it rains" was the proposed solution.

One of the frustrated environmental scientists looking for a way to get some action in Missoula was Dr. Clarence C. Gordon, professor of botany at the University of Montana. He came to visit us and convinced us to "sue the bastards" in Missoula, and soon Vic Yannacone was on his way to Missoula. Clancy had all the science well in hand, so we needed no additional input, and before long Clancy became a member of EDF's board of trustees. Hoerner-Waldorf was the only pollution source in the Missoula Valley, so there was no question who was making the mess.

Vic provided some colorful language to fire up the troops. "The effluent of the affluent, pure crud! The rape of the West! You can't see the purple mountain's majesty, much less the fruited plain. What a putrescent excrescence on the face of the globe," said Vic. On November 13, 1968, EDF became the plaintiff against Hoerner-Waldorf, requesting the pulp mill be prohibited "from depriving citizens of the Missoula ecosystem of their natural right to air unsullied by dangerous and unpleasant pollutants." Our constitutional right to life, liberty, and property was asserted, and a clean and healthful environment was our property. The case moved very slowly, and Hoerner-Waldorf made motions to dismiss on the usual grounds of lack of standing. Meanwhile, as a result of pressures aroused by the demonstrations and litigation, the company undertook a $13.5 million air pollution abatement construction program, which was well on its way to solving the problem.

Then, as had become the norm, the judge threw us out of court, case dismissed. We were becoming accustomed to that. In doing so, however, he said he had "no difficulty in finding that the rights to life and liberty and property are constitutionally protected ... and surely a person's health is what, in a most significant degree, sustains life. So it seems to me that each of us is constitutionally protected in our natural and personal state of life and health." EDF decided not to appeal the decision: We had won most of what we were after, with some excellent language besides. *Once again, we had won while losing.*

How Did Your Gasoline Get to Be "Unleaded"?

It was during those lean early days of 1969 that Dr. Paul P. Craig, a physicist from Brookhaven National Laboratory on Long Island, began attending our almost-weekly EDF board meetings. We also had nightly meetings on the telephone. At that point all of us were birders, which seemed to be a kind of prerequisite for membership in the club. Paul was not, but he was fun, smart, and dedicated to environmental protection, so we let him attend in spite of his deficiency; furthermore, we didn't have a physicist.

We were fighting the DDT wars, so Paul wanted a problem of his own. He lit on the toxicity of lead and how lead was finding its way into the human environment with serious and toxic results. Paul proceeded to become an expert on the toxic effects of lead.

Unlike the DDT issue, a very recent development, humans have been poisoning themselves with lead for thousands of years. Silver became valuable for its use in coinage, and silver mining always involved lead as a byproduct, with toxic results for miners. According to some historians, the Roman Empire disintegrated partly because of lead pipes, while the elite stored their wines in lead containers, poisoning Rome's leaders. People don't always learn from history; there always seemed to be another use for that lead byproduct.

In the 1920s tetraethyl lead was found to increase the octane rating and burning characteristics of gasoline, and from then on it was routinely added to gasoline. The permit for this additive was temporary, but that was good enough for 60 years. For decades lead oxide spewed from the exhaust of motor vehicles, contaminating landscapes and human bodies. Not surprisingly, contamination with lead was greatest near highways and included inner-city areas, frequently the homes of minorities. Lead is a neurotoxin, especially for children, and it was estimated that countless American children had lost some IQ points because of lead toxicity.

On May 5, 1970, EDF submitted a legal petition to the U.S. Department of Health, Education and Welfare (HEW) seeking reduction and elimination of lead from the exhaust of motor vehicles. The petition was written by Washington, DC, attorney Edward Berlin with scientific support by Paul Craig (Craig & Berlin, 1971). HEW cooperated with the petition, so no appeal was taken to court. Paul was appointed to an HEW committee to develop new standards for lead. EDF filed a petition with the California Air Resources Board seeking tighter standards for lead, and New York City and Washington, DC, were requested to use unleaded gasoline in their vehicles.

Progress was painfully slow, but soon federal government vehicles began using unleaded gasoline, and use of leaded gasoline slowly declined. The Ethyl Corporation and the auto industry predictably fought to preserve the fuel additive. In the 1980s getting the lead out of gasoline became a large EDF program with help from many sources and other organizations. Lead in gasoline was not only a human health hazard, but it also interfered with the catalytic converters required to meet new emission standards under the Clean Air Act. By 1987 all gasoline in the United States was unleaded.

More recent research on lead in gasoline has revealed that not only did lead contamination reduce IQ by several points in millions of children, but brain damage in early childhood led to increased aggressiveness and criminal behavior later in life. The correlations are striking. Blood lead concentrations and mental retardation in schoolchildren both peaked around 1980, while violent crime peaked about 20 years later (Drum, 2013). IQs rose and mental retardation and crime incidence declined dramatically after lead was no longer added to gasoline (Carpenter & Nevin, 2010; Needleman, 2000).

Paul Craig and EDF started the process in 1970 that ultimately made gasoline unleaded. Many people and organizations became involved in that struggle before the lead came out. *Next time you drive up to the pump and see that "unleaded" sign, remember how it got that way.*

Florissant Fossil Beds in Colorado

In May 1969, EDF received a call from Friends of Florissant regarding the threat of development and degradation of the Florissant fossil beds west of Colorado Springs. The Friends had been working for two years to save this extremely rich paleontological site, but it was about to be developed for summer homes. In July Vic Yannacone took the case to court, seeking a restraining order. The order was dismissed, so he took it to an appellate court and obtained a 17-day restraining order. A hearing and further court actions bought enough time for Congress to pass a bill and President Nixon to sign it, creating the Florissant Fossil Beds National Monument. The fossil beds were saved in perpetuity! Dr. Estella B. Leopold has written a book about this story (Leopold &

Meyer, 2012), also 40 years after the fact, like this book about DDT. Here was another early example of the value of taking an environmental issue to court.

Cross Florida Barge Canal Becomes a Greenway

In July 1969 EDF became involved with efforts to stop construction of the Cross Florida Barge Canal, a huge $210 million project to connect the Gulf of Mexico with the Atlantic Ocean. Begun in the 1930s, during World War II it was seen as a way to protect shipping from German submarines. Other excuses, such as recreation, came later. It would cause substantial environmental damage, destroy the wild and scenic Ocklawaha River, and create vast algae-choked impoundments. Several of us went to Florida to meet with scientists Drs. David S. Anthony and George W. Cornwell from the University of Florida and Marjorie H. Carr, who was spearheading the effort.

We cruised the Ocklawaha, which was indeed beautiful with original bottomland forest hanging with Spanish moss, extensive marshes, and many water birds. We plotted strategy together, helped form Florida Defenders of the Environment (FDE, our favorite three letters), and on September 15, 1969, EDF along with FDE filed suit against the U.S. Army Corps of Engineers in Federal District Court in Washington, DC. Vic Yannacone filed the action, which was later assumed by Lee Rogers, our new general counsel. It sought an injunction to stop construction pending a thorough environmental study.

FDE had conducted a substantial public relations effort for several years to save the Ocklawaha, which played a critical role in this complex case. Following oral argument, in January 1971, EDF and FDE won a significant court victory. The judge denied the Corps' motion to dismiss and issued an injunction against further construction. Four days later President Nixon stopped construction permanently. One of the nation's largest pork barrel construction projects morphed into what is known today as the Marjorie Harris Carr Cross Florida Greenway, a 110-mile corridor of natural habitats, trails, and recreation areas. Here was a case where *FDE litigation, with some help from EDF, had won by winning.*

THE ENVIRONMENTAL DECADE: THE 1970S

The new decade began with a new law, the National Environmental Policy Act (NEPA) of 1969, which required an environmental impact statement for all major federal actions. In March 1970, EDF launched its public membership, and in April 1970 the first Earth Day arrived. EDF's finances began to improve. In addition to the DDT efforts that had moved to Washington, DC, EDF initiated an increasingly varied agenda testing the limits of environmental law as the new decade began.

"And God Created Great Whales"

As new trustees or staff arrived at EDF, they tended to bring new programs with them. So it was with our new trustee, Dr. Roger S. Payne of Rockefeller University, who was an expert on whales, especially their communications. Roger teamed up with Scott McVay of Princeton University and EDF's executive director Rod Cameron, their goal being to save the world's great whales. More whales were killed during the 1960s than had died during any previous decade. Early in 1970 the EDF team tried to persuade Japan and Russia, which accounted for 85% of the world's whaling kills, to cease their whaling. In response to a formal request prepared by Rod Cameron and Scott McVay, the U.S. Bureau of Sport Fisheries and Wildlife added eight species of great whales to the endangered species list, prohibiting importation of whale products from those species into the United States. It also committed the U.S. Government to pressure the International Whaling Commission to prohibit the continued killing of those listed whale species. Many organizations, including EDF, worked on the whaling problem for years, leading to a partial moratorium on whaling established by the International Whaling Commission in 1982.

EDF Challenges Big Dam Projects

NEPA provided a new mechanism for challenging large construction projects, with standing before courts, often improving the projects before they went forward. Joining with local groups, EDF filed suit against the U.S. Army Corps of Engineers on October 1, 1970, against their proposed Gillham Dam on the Cossatot River in Arkansas. Richard S. Arnold, attorney for EDF, argued that the benefit/cost analysis of this project was greatly exaggerated and that a wild and scenic river system would be destroyed. The judge agreed and an injunction was issued. The environmental impact statement was revised, the injunction was dissolved, the Gillham Dam was built, and later the Cossatot River State Park-Natural Area was established.

With improving finances, in 1971 EDF was able to hire Dr. Leo M. Eisel, a water resources engineer with a PhD from Harvard University. Leo's economic analyses on several of EDF's dam cases clearly showed that these projects were not justified economically. The alleged benefits, often artificially inflated, were exceeded by their costs, while environmental costs were often ignored.

The Tellico Dam on the Little Tennessee River, a proposed project of the Tennessee Valley Authority, was challenged under NEPA by an EDF lawsuit in August 1971. That famous little fish, the snail darter, escalated the case under the Endangered Species Act (ESA), and the case went all the way to the U.S. Supreme Court. The Court ruled by 6–3 that the ESA prevented construction of the dam. Then Congress exempted the Tellico Dam from the ESA, and the dam was completed. The snail darter was transplanted to nearby streams and still lives.

The Tennessee-Tombigbee Waterway was a massive 253-mile-long waterway proposed to connect the Tennessee, Ohio, and Missouri rivers with the Gulf of Mexico. Proposed in 1946 to "unite the Heartland of America with the Southland," a function already provided by the Mississippi River, its benefit/cost ratio was greatly exaggerated and it would do substantial environmental damage. On July 14, 1971, EDF filed suit under NEPA, with Richard S. Arnold as attorney.

Agreeing that the U.S. Army Corps of Engineers had not followed NEPA, the judge issued an injunction against the project. After much controversy and modification, the injunction was dissolved. The project went forward and was completed in 1984.

Cache River Channelization

In 1972 EDF joined a duck-hunting dentist named Rex Hancock and other organizations in a lawsuit against the U.S. Army Corps of Engineers to prevent the continued channelization of the Cache River in Arkansas, the largest bottomland hardwood forest in the United States. EDF was represented by attorneys Richard Arnold and James T. B. Tripp, who joined EDF in 1973. On March 8, 1973, a federal judge halted the project, which was restarted in 1977 but stopped again in 1978. Finally, in 1984, the remaining intact river system became the Cache River National Wildlife Refuge. That is where the presumably extinct ivory-billed woodpecker might have been sighted in 2004–2005. Now the Corps is restoring part of the river to its original state, reversing the damage they did before. Jim Tripp is still on EDF's legal staff after 41 years.

By 1974 EDF had brought 10 lawsuits against various dams and channelization projects under NEPA, in all cases to reduce environmental damage. Courts ruled that NEPA was mostly a procedural act, where environmental effects had to be outlined in environmental impact statements, but that projects could then go forward. The Army Corps of Engineers, along with local construction and real estate interests, usually favored such projects to gain "free" federal funding in the name of flood control. As such projects continued to be built at great cost, so also did flood damage continue to increase.

We had discovered that water usually runs downhill, collecting in low areas, so flood-prone areas are entirely predictable. EDF therefore advocated flood plain management, especially not building flood-sensitive structures within the flood plain. EDF even produced a movie about this strategy: *Planning for Floods.* The wave of litigation by EDF and other organizations helped generate much-improved federal flood-control policies in future years.

Trans-Alaska Pipeline System

In March 1970, EDF, almost coincidental with launching its public membership, joined the Wilderness Society and Friends of the Earth in a lawsuit against the U.S. Secretary of the Interior to enjoin him from granting permits for the construction of the Trans-Alaska Pipeline System and its service road across the north slope of Alaska. Attorneys from the Center for Law and Social Policy in Washington, DC, handled the case, along with John Dienelt as co-counsel for EDF. (We will introduce John on page 133.) The suit sought various environmental safeguards and compliance with NEPA.

EDF argued that the pipeline to Valdez crossed dangerous seismic zones and that loading the oil into giant supertankers in a region with serious and rocky navigational problems in a stormy area teeming with marine wildlife was fraught with hazards. EDF favored a far safer overland route through Canada to Edmonton, with the oil then delivered to the American Midwest, where the need was greater than on the West Coast. Such a route could also have accommodated the already-planned natural gas pipeline.

The court issued an injunction blocking the project, and Interior then spent two years preparing a massive environmental impact statement. Its science was excellent, but it still favored the hazardous pipeline route to Valdez as chosen by the oil industry. The injunction was dissolved, only to be reinstated on appeal by the DC Court of Appeals. Then, in the summer of 1973, Congress in its wisdom exempted the pipeline system from the requirements of NEPA, the Supreme Court declined to hear the case, President Nixon signed the law, and the pipeline was ultimately built to Valdez.

Sixteen years later, on March 24, 1989, the *Exxon Valdez* ran aground, spilling between 11 and 32 million gallons of crude oil into Prince William Sound. Up to half a million seabirds, 2,800 sea otters, 250 Bald Eagles, plus seals, whales, and millions of salmon, their eggs, and fry were killed. The livelihoods of thousands of fishermen and people in related industries were destroyed. More than 20 years later oil still remains inches below the surface along 1,500 miles of Alaskan coastline. *If the fledgling EDF had prevailed in its lawsuit in the early 1970s, this catastrophic oil spill would not have happened.*

The *Exxon Valdez* suffered re-incarnation with a series of new identities: first it was the *Exxon Mediterranean*, then the *Sea River Mediterranean* and then it became the *Dong Fang Ocean* after being sold to a Hong Kong Company. Its final name was *Oriental Nicety*. It carried oil around the oceans of the world for another decade, then it carried ore, suffered a major collision and was finally sold as scrap in 2011 and dismantled in India. Oil is still being loaded into supertankers in Valdez, Alaska.

Energy Policies and Electric Utilities

In 1971 EDF hired Ernst R. Habicht Jr., a chemist with a PhD from Stanford University. He was nicknamed "Hasty" because he has always been in such a hurry that he was almost born in the back seat of his parents' 1936 Dodge. He decided he wanted to work on "energy," and after a few months he became an authority on the electricity pricing systems of utilities.

Instead of tackling pollution from electric utilities directly, Hasty led EDF to challenge the rate structure whereby electricity was sold. Since utilities at that time charged less per kilowatt-hour as consumption increased (bulk discounts), this rate structure promoted and rewarded greater consumption of electricity. Rates were the same, whether at peak or slack periods of usage. Utilities had to have the capacity to meet the peak periods of demand, with idle capacity the rest of the time. Peak usage times were the most costly for the utilities and caused the most pollution, and yet consumers who caused the peak demand paid no more than did other consumers who did not create the peak. The results were inefficient and excessive use of energy, increased pollution, decreased earnings for utilities, and inequitable distribution of costs to consumers.

Hasty and EDF, with attorney Edward Berlin and expert economists Dr. Charles J. Cicchetti and William J. Gillen, intervened in several rate-setting cases before public service commissions in New York, Michigan, and Wisconsin. They argued for time-of-day or peak-load pricing to solve these problems. Some consumers would shift their demand to off-peak times to save money, the peak would be lower, pollution would be less, and utilities would not have to build as many power plants, or enlist the most expensive energy sources, to meet the lower peak. Electricity demand would be more evenly spread, lowering the peaks and raising the valleys. Hasty's approach taught us a new term: price elasticity, which means that people buy less at higher prices, more at lower prices. These EDF arguments were strongly influenced by Dr. William S. Vickrey of Columbia University, who won the Nobel Prize for Economics in 1996, just three days prior to his death. Vickrey is best known for his advocacy of peak-load pricing to relieve traffic congestion, a strategy directly related to EDF's peak-load pricing approach for electricity.

You might think that since most players gain by such a rate modification, utilities would leap at its adoption. Not exactly! EDF had to fight for it. After three years, EDF won a major victory in 1974 when the Wisconsin Public Service Commission directed the Madison Gas and Electric Company to adopt peak-load pricing. In the 40 years since then, peak-load electricity pricing has spread around the country, especially for large consumers, but its adoption is still not as extensive as might be expected. Could it be that the coal industry does not want to sell less coal for electricity generation for the same reason that fuel efficiency in automobiles did not improve for 30 years because that would sell less oil?

This beginning in energy policy, starting in 1971, has had a profound influence on EDF's policies to the present day. First, economics became a vital discipline to EDF actions. We encountered economic arguments from opponents, so if we were to prevail, we needed economists to make economic arguments of our own. If an industry found it profitable to pollute, we would rearrange the economics to make it profitable *not to pollute*. EDF has successfully employed this strategy in many ways for the past 40 years.

Energy policies and economics have been central to numerous EDF actions. Acid rain, an energy artifact, became a major effort of the 1980s, culminating in EDF's cap-and-trade program for sulfur dioxide. It was adopted by the Clean Air Act Amendments of 1990, which resulted in a dramatic reduction in acid rain. Climate change resulting mainly from burning fossil fuels as energy sources is now EDF's largest program and the world's greatest environmental problem. Applying the successful cap-and-trade strategy to carbon dioxide emissions has been a major emphasis in EDF's climate change program.

ENVIRONMENTAL DEFENSE FUND: BIGGEST BANG FOR THE BUCK

The goal of this book has been to trace the DDT issue and its interconnections with the development of EDF through the early 1970s, but generally not beyond. The above seemingly unrelated issues show that EDF quickly took advantage of the DDT strategies to tackle new, large, and diverse problems involving land and water resources, wildlife habitat, human health hazards, and an array of energy issues. These actions were initiated when EDF was barely a fledgling organization. All it took was for the science to be organized and articulated and the package presented by competent attorneys, and the matter was then on the public agenda.

Although EDF launched its public membership in March 1970 and began to grow thereafter, total expenditures remained remarkably small for the complex of major actions that were pending. By the end of 1972, the first year that EDF produced an annual report, EDF was pursuing 80 cases around the country. For the year ending October 31, 1972, total expenditures were $677,994, an impressive example of much "bang for the buck." Less than two years earlier, EDF had arisen from near oblivion when its first three staff members were hired with only a few months' salary in the bank. Public membership rose from zero to 36,000 in those 22 months.

EDF was already showing signs of moving beyond purely a "sue the bastards" philosophy by working with governments or committees and helping develop local organizations. Economics had already appeared as a third leg on the stool, a discipline that was to become a vital force in EDF's future. "If it works, try it" was the forerunner of EDF's current tagline, "Finding the ways that work."

THE JUDICIAL APPROACH IS CONTAGIOUS

By the late 1960s it was becoming widely apparent that taking environmental issues to court was a viable strategy for environmental protection. During 1966 David A. R. Williams, an attorney from New Zealand on a postgraduate LLM program at Harvard Law School, became fascinated by EDF's approach through the courts. He visited us in our Long Island farmhouse "headquarters" where we had a spirited discussion, then returned to New Zealand. A few years later, in April 1971, he, along with a number of scientists and attorneys based in Auckland, incorporated the Environmental Defence Society (EDS). Its name and the method of approaching environmental protection came directly from EDF.

Early in 1972, I was on a lecture tour, and on several occasions I met with the directors of EDS in Auckland. "We do not have standing to sue the Crown" was a theme that came up repeatedly in our discussions, and my repeated response was "Do it anyway; just keep suing your government." EDS did that, and in a few years the standing rules had been broken down in New Zealand. Environmental groups such as EDS then had legal status in the courts. In the 40 years since then, EDS and EDF have cooperated on many international projects and problems and EDS has become an important institution for environmental protection in New Zealand.

Meanwhile in the United States, five young lawyers, four from the Yale Law School, founded the Natural Resources Defense Council (NRDC) in 1970 as a public-interest law firm. In the ensuing years NRDC has grown and thrived, along with EDF, and today it is one of the most influential of the nation's environmental advocacy organizations. Similarly, in 1971 the Sierra Club Legal Defense Fund was formed as a public-interest law firm. Its name was changed to Earthjustice in 1997. Although they are competitive in some ways, these nonprofit environmental organizations cooperate widely on many issues while dividing other issues appropriately according to their interests and talents. This informal arrangement works well and has had a significant influence on American environmental policies.

7

Time to Go After the Feds

By the fall of 1969 we knew we had to challenge pesticide regulation by the federal government if we were to ultimately prevail against DDT, but we did not know how to do it. We had the science well in hand and knew how to present it, with literally hundreds of scientists prepared to testify within their areas of expertise. We did not have the organizational structure to launch such an effort at the federal level, however, and we were certainly short of money.

At about that time Joseph L. Sax, then the leading proponent of the development of environmental law at the University of Michigan Law School, suggested that we contact the newly founded Center for Law and Social Policy (CLASP), a public-interest law firm in Washington, DC. Joe was a member of the CLASP board. He insisted that DDT was in violation of the Federal Insecticide, Fungicide and Rodenticide Act (FIFRA) and that the U.S. Department of Agriculture (USDA) was not enforcing FIFRA. I therefore called and talked at length with James W. Moorman (Fig. 7.1), attorney for

Figure 7.1 James W. Moorman, lead attorney
in *Environmental Defense Fund vs. U.S.
Department of Agriculture*, October 31, 1969.
Photo about 1978, with his permission.

CLASP, describing the DDT problem and proposed action against USDA.
"If we are going to do this, then you are going to come down here and help
me put the case together," said Jim firmly. That was not music to my ears: I
had other things I needed to do, but shortly I was on my way to Washington.

CLASP was in a rundown part of Washington, and my "housing" consisted of sleeping on an old mattress in their dusty attic. But we got to
work and wrote a petition to USDA in about a week. The petition was a
formal legal request that the FIFRA registrations for DDT be canceled.
The petition also requested that USDA suspend the registrations while it
was considering their cancellation. We had no illusions that USDA would
grant our request, but it was Jim's advice that we go to USDA for administrative relief before seeking cancellation and suspension from the courts.
In legalese, he indicated we were required to "exhaust our administrative
remedies" before seeking "judicial review." If, as we anticipated, USDA
rejected our requests, we planned to seek review from the U.S. Court of
Appeals for the District of Columbia Circuit, which Jim indicated was the
appropriate court in which to seek review.

Jim wrote the legal petition and I wrote the science part as an accompanying affidavit. The science was much like our case in Madison. Our petition was supported by 70 reprinted scientific papers with 268 references. We were getting better and better at this with each step of the way, with new information arriving almost weekly. The petition had been very carefully drafted by Jim Moorman with significant assistance from Joe Sax, Charles Halpern, Bruce Terris, and legal interns at CLASP. They understood the intricacies of the standing problem and the determination of the Justice Department (defending USDA) to see that we would be dismissed. None of us with EDF fully understood the legal impediments, although we were beginning to realize that our case would not be heard by a federal agency or court unless the standing barrier was overcome. The science seemed easy by comparison.

We filed our petition on October 31, 1969, asking the Secretary of Agriculture, Clifford M. Hardin, to immediately bar the use of DDT by suspending and canceling its registrations as a pesticide. In addition to EDF, which still had no office, staff, or public membership, the petition was filed on behalf of the Sierra Club, the National Audubon Society, and the West Michigan Environmental Action Council. Former Secretary of the Interior Stewart L. Udall announced the filing at a news conference in Washington that morning, describing DDT as "the uninvited additive!" Udall had become a strong supporter and before long he joined EDF's board of trustees as an active member. *Here was a genuine public servant, dedicated to the public interest and environmental protection.*

The title of our petition was "Petition Requesting the Suspension and Cancellation of Registration of Economic Poisons Containing DDT." It set out the reasons why DDT was improperly registered under FIFRA. FIFRA is an odd statute in that its substantive requirements are on the pesticide label. Thus, FIFRA provides for a label "which may be necessary and, if complied with, adequate to prevent injury to living man and other vertebrate animals, vegetation, and useful invertebrate animals." If, when used according to the label, a pesticide does not protect the specified nontarget organisms, then it is termed "misbranded" and its registration should be "canceled." Our petition argued that DDT once applied was a persistent and uncontrolled biocide in the environment that injured non-target organisms and that there was no label that could be written that would protect nontarget organisms. DDT therefore was "misbranded" and its registrations must be canceled.

The cancellation procedure had never been applied by USDA under FIFRA, except when requested by the manufacturer, so the procedure had never been tested. We were concerned that the undefined process of cancellation could involve a lengthy delay during which DDT would remain on the market and in use. For that reason, our petition asked for immediate suspension of all DDT registrations pending a final decision on cancellation.

When a pesticide is used according to the label, yet nontarget organisms are damaged nevertheless, and when no label change can protect nontarget organisms from damage, then the pesticide is termed an "imminent hazard to the public," and FIFRA provides for its suspension. Suspension takes the product off the market immediately while cancellation procedures are implemented. In our petition we argued that because DDT was moving about the world contaminating and damaging countless nontarget organisms (including humanity) and ecosystems, there was no label that could provide protection, and thus DDT should be both canceled and suspended.

The EDF petition also included alternative pest control techniques, including both chemical and nonchemical procedures, that could control insect pest problems without the use of DDT. EDF has always offered a better way to solve a problem while avoiding the practice that did environmental damage. Here was the beginning of the evolution from "sue the bastards" to "finding the ways that work," EDF's tagline. It's always easier to win a battle when you not only want something stopped, but a better alternative started.

WHAT!? DDT ALSO CAUSES CANCER?

During the latter part of 1969 we became aware of data showing that DDT causes cancer in mice and rats. At first we dismissed this information, not appreciating the significance of tumor causation in rodents. Furthermore, we neither needed nor wanted it. We believed we had enough information to remove DDT from the market because of wildlife damage. We wanted this to be a wildlife case, to win on behalf of wildlife protection, without the additional burden of having to prove human health damage to protect wildlife. But we could not ignore these data.

This meant that we had to expand our science expertise to include chemical carcinogenesis and needed experts in that field. We did so, contacting world-class experts Drs. Samuel S. Epstein at Case Western Reserve University and Marvin A. Schneiderman and Umberto Saffiotti of the National Cancer Institute. They made it clear to us that tumor formation in carefully controlled rodent tests sponsored by the National Cancer Institute indicated DDT to be a cancer hazard to humans. Therefore, on October 7, 1969, before the USDA petition was filed, we filed another petition with the U.S. Department of Health, Education, and Welfare (HEW) alleging that DDT was a cancer danger to humans. This petition was written by Edward Berlin of the Washington, DC, law firm of Berlin, Roisman, and Kessler.

The Delaney Food Additives Amendment of the Food, Drug, and Cosmetic Act stated

> That no additive shall be deemed to be safe if it is found to induce cancer when ingested by man or animal, or if it is found, after tests which are appropriate for the evaluation of the safety of food additives, to induce cancer in man or animal.

DDT had become an unintended or "uninvited" food additive, and our petition claimed that HEW was obligated under that act to establish a zero tolerance for DDT in food products. The Delaney clause of the act correctly presumes that there is no safe threshold for carcinogens in human foods, and any amount of carcinogens must therefore be prohibited from entering the food supply. Our petitions under both FIFRA and the Food, Drug, and Cosmetic Act were therefore entirely consistent with the protections afforded by law existing at that time.

WHAT!? DDT ALSO CAUSES MUTATIONS AND BIRTH DEFECTS?

Yes, mutations and birth defects as well! Recent studies from the Food and Drug Administration (FDA) had demonstrated that DDT causes mutations in rats. Mutagenesis and carcinogenesis in test animals indicate a high probability, but not a certainty, that DDT causes mutations and cancer in man. Some of its deleterious effects may therefore appear in future generations.

Tests for mutagenesis and carcinogenesis cannot be performed on human subjects for many reasons, including moral ones. Instead, laboratory animals, usually rodents, are tested at high dosage levels. Relatively few agents are capable of producing these rather specific forms of genetic damage. Most of the chlorinated hydrocarbon insecticides have this capability, which we later documented at the DDT cancellation hearing in Washington, DC.

BOTH FEDERAL AGENCIES REACT TO THE EDF PETITIONS. BARELY!

The responses of both federal agencies, USDA and HEW, were underwhelming! On November 20, 1969, USDA issued notices of cancellation of DDT registrations for the following uses: (1) on shade trees (including elms for Dutch elm disease control); (2) on tobacco; (3) in and around the home; and (4) in aquatic environments, marshes, wetlands, and adjacent areas, except those that are essential for the control of disease vectors as determined by public health officials. EDF's actions had already reduced or eliminated some of these uses in various states.

We were unimpressed. By far the largest use of DDT was on cotton, which was unaffected. Furthermore, by not suspending the canceled registrations, DDT could still be bought in bulk for both canceled and uncanceled registrations and then used anywhere while the administrative cancellation procedures dragged on. The heretofore never-used cancellation process presumably would involve a sequence of committees, studies, reports, and hearings that could take years, all while the harmful use of DDT continued. The manufacturers of DDT were already contesting the token cancellations of DDT made by USDA in response to our petition, thereby nullifying the minor potential positive effects of even this action, which was really no more than a change of labels. By canceling but not suspending a few uses of DDT, USDA apparently sought to convince the public that DDT usage had been greatly reduced. The reality was that nothing had happened; the use of DDT was to continue as before.

In addition, USDA provided a second review procedure in which "interested persons" could "submit views and comments on this proposal" for 90 days. USDA was giving industry extra time to build a record. Pro-DDT articles were already appearing in the public media, and we presumed that USDA was being deluged with pro-DDT material from industry. At the end of 90 days, USDA could announce that it had received far more evidence in favor of DDT than against it, and that, under the circumstances, DDT applications could continue.

Of crucial importance to EDF and all environmental citizen organizations, USDA contended that environmental protection groups have no place in the decision-making process concerning pesticides. USDA took the position that EDF and its allies did not have standing to seek the cancellation or suspension of DDT registrations. In USDA's view, only USDA and the pesticide companies should be left to safeguard the nation's environment. This was a direct challenge to our plan to take USDA's denial of our petition to the DC Appeals Court for review. USDA was hiding behind the standing barrier to block public input, to prevent a public forum from exposing the science supporting the DDT threat, and to continue doing business as usual. Unless the standing barrier was overcome, our case would be shunted aside.

On November 12, 1969, HEW also denied our petition, announcing "that the use of DDT . . . be restricted within two years to those uses essential to the preservation of human health and welfare." Details were not given and the word "essential" was not defined. The FDA, acting on behalf of HEW, did not dispute the scientific evidence that DDT is a carcinogen and an environmental hazard generally. Instead, they suggested that "it is not practical to attempt to eliminate the residues of persistent pesticides from food by the establishment of zero tolerance limits." Our petition had not asked for that, however, instead requesting that future additions of DDT to foods be stopped by a suspension order.

THE ENVIRONMENTAL DEFENSE FUND TAKES TWO FEDERAL AGENCIES TO COURT

Since both HEW and USDA had substantially denied our petitions, the next step by EDF to escalate the DDT battle was to seek judicial review. Accordingly, a petition for review was filed in the U.S. Court of Appeals for the District of Columbia Circuit, the second highest court in the land. The stakes for the future of the DDT issue, and of EDF as an organization, were getting ever larger. Our appeal to the DC Appeals Court was filed on December 29, 1969, seeking more effective and swifter action to eliminate the use of DDT than had been taken so far.

Our USDA case, in effect, requested the court to order USDA to issue notices of cancellation for all economic poisons that contain DDT and that all such registrations be suspended immediately. An underlying assumption of our court action was that EDF and co-petitioners would be appropriate participants in the cancellation proceedings. Attorneys Jim Moorman and Charles Halpern prepared EDF's brief filed in support of the USDA case.

In the brief to the court prepared by Ed Berlin regarding our HEW petition, we asserted that HEW did not deny that "DDT causes cancer in test animals and is responsible for widespread environmental degradation, including damage to non-target organisms of many species, the preservation of which is essential to the well being of man." In response to this evidence, the secretary of HEW not only refused to take immediate action to prohibit the further contamination of raw agricultural commodities with DDT, but also refused to even publish notice of our proposal in the *Federal Register*, thereby denying us access to the administrative procedures available under the Food, Drug, and Cosmetic Act.

I hope readers will pardon the legalese, but it takes a lot of verbiage and quotations from the acts to justify these actions and bring them before the appellate court. Our USDA appeal was a small book of 38 pages with 60 pages of addendums. The HEW appeal was 40 pages with 20 pages of supplements. The battle against DDT was getting bigger and more complex, and promised to become more so. All this was happening with the driving force coming from volunteer scientists and other citizens fighting the federal government, most of them with other jobs and responsibilities—a true grassroots movement if ever there was one. Nobody was paying us to force an evaluation of the DDT problem, a job that should have been performed by the two relevant federal agencies, USDA and HEW. Making them do their jobs was like pulling teeth. We were trying to force them to act under the laws they supposedly were administering.

THE APPEALS COURT MAKES AN HISTORIC DECISION

"consumers . . . have standing to protect the public interest . . . "
—CHIEF JUDGE DAVID L. BAZELON

Oral argument by attorney Ed Berlin for the HEW case took place on April 10, 1970. The USDA case was not yet ready for oral argument on the merits because the U.S. Department of Justice had filed a vigorous motion to dismiss the case for lack of standing by EDF and its allies to seek judicial review. Following yet more briefs by Jim and his colleagues at CLASP on behalf of EDF, the appeals court decided to hear the case on its merits.

On May 28, 1970, the U.S. Court of Appeals for the District of Columbia Circuit rendered two major decisions against HEW and USDA establishing important precedents in environmental law. First, the court ordered HEW to publish EDF's petition in the *Federal Register*, which started an administrative process to consider elimination of tolerances for DDT; HEW had said that doing so was impractical. USDA had argued that EDF lacked standing to bring the action, that the court lacked jurisdiction, that EDF was in the wrong court, and that USDA's decision was not final and therefore not yet reviewable. Chief Judge David L. Bazelon, writing for the court, rejected every one of those contentions. He stated that

> *the consumers' interest in environmental protection may properly be represented by a membership association with an organizational interest in the problem . . . [C]onsumers . . . have standing to protect the public interest in the proper administration of a regulatory system enacted for their benefit. The interest . . . need not be economic.*

Whereas USDA had claimed that only the registrant (manufacturer) of the pesticide could appeal a decision regarding pesticide registration, the court ruled that "the 'zone of interests' . . . protected by the statute includes not only the economic interest of the registrant but also the interest of the public in safety." The court called USDA's actions against DDT "a few feeble gestures." "The controversy is as ripe for judicial consideration as it can ever be," said the judge, adding that USDA's "decision is not . . . placed beyond judicial scrutiny." "If he [the Secretary of Agriculture] persists in denying suspension in the face of the impressive evidence presented by petitioners [EDF et al.], then the basis for that decision should appear clearly on the record."

This was a remarkable court victory. Our original strategy that someday, some court, somewhere, would open the locked courthouse door and let us in had borne fruit. *Indeed, it was the second highest court in the land, and our case was against two huge federal agencies. The standing barrier preventing citizens from suing the government had been weakened, if not broken.* In addition, the court ordered USDA to suspend DDT, or to give the reasons for not suspending. Our DDT battles were certainly not over, but we had taken a large stride forward: *two huge Federal agencies, USDA and HEW had been scolded by the court.*

As the court rendered these national precedent-setting decisions, EDF had opened a small office above the Stony Brook post office and had three employees and its first few members after launching a public membership drive. Viewed from today's perspective, one might marvel at the initiative of this court to protect the environment and public health: Would today's courts, not to mention Congress, do as well?

THE BRUISING BATTLE OF THE BRIEFS

In its precedent-setting decision of May 28, 1970, the appeals court had ordered USDA to either suspend all registrations of DDT within 30 days or give the court its reasons for failing to do so. USDA chose the latter. Supported by its response of June 29, 1970, which EDF considered incompetent and largely irrelevant, USDA stood firm on all registrations. Then on August 10, CLASP filed a lengthy reply brief with the court on EDF's behalf. USDA filed another brief on August 31, and EDF responded with another but shorter reply brief on September 8. Oral argument by Jim Moorman was heard on September 9. The fate of DDT in the United States was then left in the hands of the court.

By this time the DDT litigation had become a major national environmental test case. In addition to the original petitioners (EDF, Sierra Club, National Audubon Society, and West Michigan Environmental Action Council), EDF was joined by the Izaak Walton League of America and the State of New York as interveners. USDA was joined by the Montrose Chemical Company (the major manufacturer of DDT) and the National Agricultural Chemicals Association, the all-too-familiar picture of an agency aligned with the industry it is supposed to be regulating.

HEW responded differently to the court order. Since the Delaney clause of the Food, Drug, and Cosmetic Act required a zero tolerance for any "food additive" found to cause cancer in experimental animals, HEW had been ordered by the court to either establish zero tolerances or explain why a greater tolerance would be safe. Since many foods were already contaminated by DDT, the court had suggested that the zero tolerance be exempted for preexisting DDT contamination. Another suggestion by EDF was to set zero tolerance as a goal to be phased in over time, with no new DDT added to the system. HEW did not seek a review of the court ruling but instead sought to comply with it.

My apologies again for burdening readers with the legal complexities of this case, but it demonstrates the immense legal thicket that existed to protect industry and its products from public scrutiny and decision making. We had to get through that thicket, especially the standing barrier, before the scientific case against DDT could be heard and before EDF could play a role in protecting the environment and public health. We thank the legal team in Washington for their brilliance and perseverance in helping create what we now call environmental law, which allows public representation on issues to be heard and decided on their merits.

8

Escalating the DDT Issue with More Court Cases

While HEW and USDA pondered these appellate court decisions, we turned our attention to several more local DDT problems. From a *New York Times* article (May 3, 1970), we learned that the Olin Chemical Corporation was manufacturing about 20% of the nation's DDT in buildings owned by the federal government and leased to Olin on the site of the U.S. Army's Redstone Arsenal near Huntsville, Alabama. A DDT-contaminated effluent from this plant was leaking into the Wheeler National Wildlife Refuge at concentrations known to inhibit reproduction of birds and fish. The refuge also served as a drinking water supply for the city of Decatur, implying a human health hazard as well. Downriver fisherman were also eating their catch, thus concentrating DDT to higher levels as well.

In October 1969, the federal Water Quality Administration had recommended a stricter pollution control standard for the Olin plant. Olin said it could not meet that standard, and the Army then overruled the

Water Quality Administration's recommendation. So on June 5, 1970, EDF, along with the National Audubon Society and the National Wildlife Federation, sued in Federal District Court against Olin, the Department of the Army, and the Corps of Engineers seeking to stop the DDT-contaminated discharge. The complaint was written by EDF's new attorney, Edward Lee Rogers. I supplied the scientific support, which was easy, since it was similar, although steadily expanding, to the Wisconsin hearings and the USDA and HEW cases.

Only three days later Olin threw in the towel! On June 8 Olin decided to close its DDT plant and no longer make DDT. DDT apparently was not worth defending. They said they had reached that decision shortly before our case was filed. True or not, it was a quick and easy victory. We needed it. *We had won by winning.*

FLUSHING TONS OF DDT INTO THE LOS ANGELES SEWER SYSTEM

Even as the legal briefs went back and forth between EDF, USDA, HEW, and the appeals court, another DDT battle was brewing in California. For years scientists had been puzzled by the extremely high levels of DDT contamination along the coast of Southern California compared with other marine environments. Fish from Santa Monica Bay carried many times the concentrations of DDT as did fish from San Francisco Bay, which drains most of California's agricultural regions. Brown Pelicans nesting on Anacapa Island in the Channel Islands were highly contaminated with DDT, there was complete nesting failure, and eggshells were so thin that the colony was littered with broken eggshells (Risebrough, Sibley, & Kirven, 1971). The birds could not reproduce in Southern California. Double-crested Cormorants also laid thin-shelled eggs and experienced nesting failure (Gress et al., 1973). Many other species of birds and fish were highly contaminated in these waters.

The Montrose Chemical Corporation, located in Los Angeles, was then the world's largest manufacturer of DDT, producing about 100 million pounds of DDT annually. The suspicion was that DDT contamination was coming from the Montrose plant. We soon learned that massive quantities of DDT, 100 to 1,000 pounds per day, were passing through the sewer line, into a treatment plant, and then into the Pacific Ocean. Upstream of Montrose DDT amounts in the sewer line were nil. Beginning in 1953, Montrose was estimated to have flushed more than 1,800 metric tons of DDT through the Los Angeles sewer system. An additional 350 to 700 metric tons of DDT in an acid sludge from the Montrose plant were dumped directly into the ocean.

Further indication of the magnitude of the Montrose pollution came from a sedentary, nonmigratory species of crab, which was analyzed for DDT residues along the West Coast from Ensenada, Mexico, to the Golden Gate Bridge off San Francisco on the north. DDT residues increased from both directions to a great peak in the marine waters off Los Angeles (Burnett, 1971). Crabs near the Los Angeles sewer outfall contained 45 times as much DDT residues as did those in areas draining California's major agricultural areas under the Golden Gate Bridge.

On October 22, 1970, EDF filed suit in Los Angeles Federal District Court to compel Montrose to stop discharging DDT into the sewer system. The complaint was written by EDF attorney Edward Lee Rogers and William H. Rodgers of the University of Washington Law School, who had become an EDF trustee by then. Scientific support was by Bob Risebrough. After an eight-month delay, EDF attorney Norman Rudman inspected the Montrose plant, which had installed equipment to prevent any further discharge, and the plant disconnected itself from the sewer system. As a result, EDF recommended to the court that the case be dismissed. It was years later that we learned the Los Angeles County sanitation districts had learned about the discharge and was insisting that it be stopped before the EDF suit was filed.

WHAT HAPPENED TO MONTROSE AND ITS POLLUTION LEGACY AFTER 1972?

The discharge of DDT through the Los Angeles sewer system was stopped in 1970 after 17 years, but Montrose continued to produce DDT for export until 1982. The plant then was closed and Montrose went into bankruptcy. Montrose ownership was intertwined with three other companies, including Chris-Craft, the boat maker, and Stauffer Chemical. In 1990 the National Oceanic and Atmospheric Administration (NOAA), along with the State of California and the Environmental Protection Agency, sued eight companies, including Montrose, for damages to the marine ecosystem, fish, and wildlife estimated at $1 billion. Fifteen sanitation districts within Los Angeles County were also named defendants in the litigation. It turned into a full decade of extremely complex litigation (http://www.darrp.noaa.gov/southwest/montrose/pdf/msrp_rp_section2.pdf).

Montrose had taken out liability insurance with seven separate insurance companies. When the insurance companies refused to defend Montrose in court, Montrose then sued all seven of its insurance companies. Montrose continued to argue that the DDT found in fish and birds was coming from agricultural runoff (remember the crab data?), not from the tons of DDT in the sediments from their discharge. Montrose further contended that the DDT residues were degrading naturally and posed little threat to marine life or people. It didn't seem to matter that this defense was absurd.

By 1999 all parties were exhausted, and a damage settlement was finally reached. The Montrose companies and the cities paid a total of $140 million, with $30 million designated for restoring fish, wildlife, and other natural resources and $33 million for containment of the sediments. The Environmental Protection Agency got the rest for Superfund cleanup.

The Montrose DDT discharges created one of the most impressive pollution sites in the country. The discharge was estimated to have been 10 times the total chlorinated hydrocarbons discharged by the Mississippi River draining the agricultural central United States each year. One hundred ten tons of DDT residues, mostly as DDE, remain spread over 17 square miles of Palos Verdes shelf sediments at depths of two inches to two feet. Small areas have been capped by clean fill, but most of this area of contamination continues to feed DDE into marine food chains. Montrose was designated a hazardous waste Superfund site and placed on the National Priorities List for cleanup by the Environmental Protection Agency.

The biological effects of this pollution source are equally impressive. Bottom-feeding white croakers and kelp bass were too contaminated to reproduce, and DDT residues in those fish exceeded federal and state guidelines. There were 20 to 30 pairs of Peregrine Falcons on the Channel Islands until 1945, but by the early 1960s they were gone. Brown Pelicans, Double-crested, Pelagic, and Brandt's Cormorants, Ashy Storm-petrels, Cassin's Auklets, Pigeon Guillemots, and Western Gulls showed eggshell thinning and/or population declines. In the early 20th century there were 35 Bald Eagle nests on the Channel Islands, but the last active nest was in 1949, and by 1960 there were none left. In 1980 Bald Eagles were reintroduced into Santa Catalina Island, but without reproductive success.

BALD EAGLES GONE, GOLDEN EAGLES MOVE IN

With Bald Eagles no longer present in the Channel Islands, Golden Eagles were attracted by feral pigs and sheep, and by the island fox, native to the islands. Golden Eagles normally feed on rabbits and other herbivores, which have short food chains, so they do not accumulate DDT like Bald Eagles, which feed on fish at the end of a long food chain contaminated by the Montrose DDT burden. Ranges of the two eagle species do not normally overlap, and Bald Eagles are dominant over Golden Eagles and drive them away. Soon the island fox population began to decline and became endangered from predation by the Golden Eagles.

Using Montrose restoration funding, efforts to save the island fox meant moving the Golden Eagles out and bringing the Bald Eagles back. By 2000, 13 Golden Eagles had been trapped and relocated to other parts of California, but the fox numbers continued to decline. Island foxes were captive-bred and released. By 2004 another 18 Golden Eagles had been relocated, but the remaining ones had become extremely wary and difficult to trap. Innovative techniques were necessary, but the last Golden Eagle was finally removed in 2006. None of the relocated birds returned. Feral pigs, cats, and exotic livestock were removed from the islands (http://iws .org/species_islandfox_SCruz.html).

In 2002 efforts began to bring back the Bald Eagles, funded by the Montrose Settlements Restoration Program. By 2006, 61 zoo-raised immature Bald Eagles were released on the Channel Islands, and in 2006 two nests each fledged a chick, the first successful Bald Eagle breeding on the islands since 1950. There was successful nesting each year since then, and at the end of 2010 there were 52 Bald Eagles on the Channel Islands.

Peregrines began recolonization naturally, which was enhanced by captive breeding and releases. Between 1978 and 1993 the Santa Cruz Predatory Bird Research Group released almost 800 Peregrine Falcons in California, 37 of them on the Channel Islands. Releases continued and by 2007 there were 25 breeding pairs of peregrines on the islands. Breeding success increased with increasing distance from the center of DDT contamination. Nests closer to the contaminated sediments had shell thinning above the threshold of about 17% associated with egg breakage. The Montrose Restoration Program also undertook extensive measures to enhance seabird populations, such as the removal of feral cats.

It is now 44 years since the Montrose DDT discharge ended. Marine food chain contamination is diminishing and Channel Islands fauna is approaching pre-DDT levels. The Montrose restoration funds helped a great deal, as did numerous volunteers who cared about the environment of Southern California. But tons of DDT residues still lie in the sediments, and they will still be there in the coming decades. Their effects are still appearing.

The California Condor has been brought back from the brink of extinction, as low as 22 birds in 1982, to more than 400 today. Extensive captive breeding and releases by The Peregrine Fund in Boise, the Los Angeles and San Diego zoos, and elsewhere are a remarkable success story. Condors have been released in several locations, including the coastal mountains near Big Sur, California. Now those birds have been found to be laying thin-shelled eggs. They have been feeding on DDE contaminated carrion sea lions along the coast. Some of those eggs have been exchanged in the nests for eggs with normal shell thickness from the San Diego Zoo (Burnett et al., 2013). Lead poisoning is an even greater problem for condors.

Dr. Samuel K. Wasser, director of the Center for Conservation Biology at the University of Washington in Seattle, has been studying killer whales in the straits off the San Juan Islands, along with his PhD student, Jessica Lundin. They use dogs trained to locate the whale scat, which floats. From the scat his team can identify many aspects of the condition of the animals, including their DDE and PCB contamination. Three fish-eating killer whale pods are resident from May through October in Washington State waters, but two of these pods spend much of the winter feeding on fish off the California coast. Those two pods have substantially higher DDE levels than the pod that remains in the north during the winter. The Montrose DDT pollution legacy lives on after more than 40 years (Krahn et al., 2007, 2009; Wasser, personal communication).

9

Changes in Washington Affect the DDT Battle

Late in 1970, President Nixon proposed and Congress approved creation of the Environmental Protection Agency (EPA), in the process transferring the Pesticide Regulation Division from USDA to EPA. For pesticide regulation, this was no minor matter. The transfer was from USDA, an agency that primarily protected pesticide manufacturers and promoted their products, to EPA, an agency that was directly charged with protecting the environment. That was to make a large difference in how the DDT issue would be resolved. The first administrator of EPA was William D. Ruckelshaus, an attorney with a sterling record of public service in government.

The other major item was the decision on DDT from the DC Court of Appeals. On January 7, 1971, the court ordered Ruckelshaus to immediately cancel all registrations of DDT and to determine whether DDT was "an imminent hazard to the public" and therefore should be suspended. The court was clearly annoyed by USDA's failure to give adequate reasons for not suspending, so "it will be necessary to remand the case once more, for a fresh determination" of the matter of suspension. The court had taken away the discretion usually afforded a federal agency and ordered it to take action. This was an unprecedented decision. EPA had only been created on December 2, 1970; Ruckelshaus barely had time to find his telephone before this court order landed on his desk as his first order of business.

Perhaps the most important part of this decision was that EDF survived USDA's motions to throw our case out of court. *The standing for citizens to sue the government, previously unavailable, had now been established by this precedent-setting decision. This was the firm beginning of what we now call "environmental law."* But you should not take the legal conclusion of a lowly scientist (me). Instead, here are the words of Joseph L. Sax, a professor of law at the University of Michigan Law School, from his September 30, 1973, letter in support of EDF's application for the Tyler Ecology Award (we did not get it). If anybody is the "father" of environmental law, it's Joe Sax (Sax, 1970). This is most, but not all, of his letter, taken from a carbon-paper copy that has been buried in my file for all these years (remember carbon paper?). Even Joe didn't remember it from nearly 40 years ago, but he gave me permission to reproduce it.

My own field is environmental law, and I have followed developments in this field quite closely for about 10 years. On the basis of my observation, I have no doubt that EDF has been the single most significant organization working in the field of environmental law. It was the first, and most innovative, group in seeking to bring the legal system to bear upon environmental problems; and it has—in respect to DDT—brought the most decisive result to fruition as a consequence of unceasing and strategically brilliant use of the legal system.

Prior to EDF's embarking upon a program of litigation to restrain the use of DDT, there was nothing that could be called, in any systematic way, a field of environmental law. There were, of course, occasional suits brought in the field of air and water pollution, and periodically cases would be initiated to challenge a public project, such as a hydroelectric dam. EDF literally invented the idea that the courts might be used effectively to get the legal system to implement ideas for environmental improvement that were abroad in the scientific community.

It is difficult to convey to anyone who has not watched the legal system change over the last decade the boldness and creativeness of this idea. I myself was frankly doubtful that the program EDF had in its collective mind was possible of achievement; a good idea and a needful one it was, but one would have been oblivious to the resistance of the legal system and the fundamental changes EDF sought to make, to have believed that it could succeed. Yet succeed it did, and it is not too much to say that the very existence of the field of environmental law today is largely its contribution.

For many years, there were only occasional breaches in the firm legal tradition that protection of the public interest must be left to public agencies. So long as that tradition held, it was not possible for citizens to play a significant role in legal protection of the environment. To put it another way, the very possibility of an environmental movement using the legal system was essentially absent until this barrier, the doctrine of standing to sue, was broken down. The EDF cases, and particularly the striking decisions they obtained against all odds, in the U.S. Court of Appeals for the District of Columbia Circuit (and that have been approved in other cases in the Supreme Court) were a central element in breaking down that barrier.

If EDF had made no other contribution to the legal system than this, they would have made a highly significant mark in our legal history. But this was only the first of many landmarks their work set out. In challenging the administration of the federal pesticide law (the so-called FIFRA), they also developed the essential precedent that a law generally describing protection for the public against harm was a law that created enforceable rights for members of the public. This may seem obvious, but the law had always been quite different. It was the traditional, and deeply rooted, rule that such protective laws did not create enforceable legal rights for members of the public, but were treated as directions to the federal agency that were to be enforced, if at all, by the Congress. Obtaining a precedent that such a law created an enforceable citizen right was a tremendous achievement, and it too has become an indispensable element of citizen action in the environmental movement. Almost all cases of recent years in the environmental area stand on the shoulders of this precedent.

Let me turn finally to their achievement. There are many environmental organizations that have taken on a case and made a useful precedent. There are, of course, instances where such organizations have obtained a particular result they sought, as with a particular highway, or a dam or wetland filling project. I know of no organization that has taken on a problem area, as EDF did with pesticides and DDT particularly, in which there has been anything like the long term commitment and sense of purpose that characterizes EDF. They moved from forum to forum and state to state, seeking an effective means to link the enforcement powers of the legal system to the scientific evidence they sought to put before decision makers. In ultimately succeeding, after long and arduous efforts in various states, before administrative agencies, in the federal courts and the Environmental Protection Agency, they have demonstrated what it means to have—in the fullest and most impressive sense of the word—a grand legal strategy. There is nothing to compare with it in the legal world during the past 15 years.

While I am not an officer or official of EDF, I have been well acquainted with the organization since its earliest days. I know the enormous difficulties it faced in obtaining even the most miniscule financial support for its first efforts. I know the reluctance of foundations to help what then seemed grandiose and even fantastic aspirations. I remember the discouraging days in Michigan, in the early cases there, where the courts could hardly believe the claims that were being made, claims of a kind they had never heard before. To have weathered those discouragements, to have struggled without pause to educate judges and administrators to the urgency of hearing and responding to scientific evidence of environmental degradation, is literally awe-inspiring. To give you some notion of what it was like in those "old days," let me quote a judge in the late 1960's when an environmental case came before him for trial. He opened the proceedings by telling the parties: "I will advise you that . . . before this case started I looked up the meaning in the dictionary of 'ecology' because I noted it in the [documents]. I was not aware of it before."

In compliance with the appellate court order, Ruckelshaus issued notices of cancellation for all registrations of DDT, but on March 18, 1971, he refused suspension. EPA seemed to be on both sides of the fence at the same time. EDF appealed EPA's failure to suspend, claiming that EPA's reasons for not suspending were inadequate, that this decision was inconsistent with their own findings of fact, and that DDT's damage to wildlife and the hazards to human health required suspension. Meanwhile we would have to go forward with the cancellation process, whatever that was going to be, while use of DDT continued. Representing EDF in this process was our new attorney, William A. Butler. It was now the summer of 1971, nearly two years after the filing of our original petitions to USDA and HEW, five years after our earliest DDT case on Long Island. But there was no doubt that we were making headway.

EDF BEGINS TO EXPAND: NEW PEOPLE, NEW WASHINGTON OFFICE

Following EDF's announcement of its public membership in March 1970, and with an assortment of actions in several arenas with much publicity, our finances began to improve. By November 1970, EDF had 10,000 members, which grew to 15,000 by March 1971. This was, after all, the "environmental era" and the first Earth Day had been launched. It felt like environmental protection and awareness was in a steady upward trajectory. EDF "headquarters" moved from the top of the Stony Brook post office to "a 100-year-old farmhouse, badly run down but still bearing traces of its former splendor" (Marion Rogers in *Acorn Days*). It was in Setauket, still on the North Shore of Long Island (Fig. 9.1). The term "headquarters" suited Bob Smolker's Mystique Committee very well, giving the impression there was a chain of offices around the country, which there was not. The old farmhouse was it.

During the summer of 1970 we hired William A. Butler, a young attorney with a BA from Stanford, a law degree from Yale, and a PhD from Harvard (Fig. 9.2). He wanted to open an EDF office in Washington, DC. In Bill's words:

> At first they [the local trustees] were a little skeptical of this idea, but . . . it apparently grew on them, particularly when they found they had enough money to risk it. So they offered me a job and, more surprising, I accepted. They could only guarantee me three months' salary [at a rate of $9,000 per annum]. But I liked the idea of being my own boss in my own office, even if the office consisted only of me.

Figure 9.1 EDF headquarters from 1970 to 1977 was this 100-year-old farmhouse in Setauket, Long Island, New York. Sketch by Marie H. Gladwish, with her permission.

Figure 9.2 Attorney Bill Butler, who represented EDF in the hearings in Washington, DC, on DDT, aldrin/dieldrin, and mirex. Photo about 1972, with his permission.

With all this activity going on in Washington, we figured maybe we'd better have an office there. We had no experience in opening offices, least of all in Washington, so we sent Bill to go do it. Bill's own words from *Acorn Days* describe best what happened:

Some of EDF's work in [Washington, DC,] had been done by the public law firm of Berlin, Roisman and Kessler. Because of this association the EDF Board graciously rented a bit of space for me in this firm's offices. I was more than a little appalled when I first saw their offices, which were in a row of ramshackle old frame houses at 1910 N Street NW. Right across the way was the law firm of Arnold and Porter, which had some very nice row houses, in contrast to our shabby heap. The reception room for [EDF] was the women's bath-room, . . . complete with bathtub, shower and toilet facility. When we took over that room we made now-U.S. District Court Judge Gladys Kessler use the men's room, something she reminds me of whenever I see her, even years later.

I spent much of the blisteringly hot August of 1970 building frames over the bathtub and the toilet with a Stanford classmate of mine who was working at [NIH] but who was also an excellent carpenter with some spare time after hours. We did not attach these frames to the wall. In the event EDF failed, we wanted to be able to return the room as quickly as possible to its original bathroom status. I hired Marie Bauman as our first part-time secretary and she generously made curtains for this room. The curtains existed until fairly recently as dust rags in the EDF DC office.

My office, directly under our reception room, down a pair of rickety stairs, was the furnace room for the building. It also housed the gas hot water heater, which not only occasionally leaked gas, which meant that I had to keep the windows open summer and winter, but also leaked water, which meant that frequently there was an inch of water on the floor. I kept a pair of overshoes in my lower desk drawer for those occasions. When we had a visitor Marie signaled me by either stomping on the floor or flushing the toilet. The rent for these palatial quarters was $50 a month, including space in my office for my bicycle.

I (CFW) was certainly impressed on my first trip to the office when Bill reached under the desk and flushed the toilet. It proved to be a good conversation starter with visitors.

10

DDT Goes to Trial, Finally, in Washington, DC

Chemicals are not "innocent until proven guilty"; they do not have human rights. Unless we treat them as guilty until proven innocent, they will deprive real human beings of their rights to health. The burden of proof must be on the chemicals and their makers—not on the human population.

—CHARLES F. WURSTER (1972)

EPA was only five weeks old on January 7, 1971, when the Court of Appeals ordered the agency to cancel all DDT registrations. The situation was fluid, to say the least. EPA did not know how the cancellation process was to be carried out, since USDA before them had never executed a cancellation procedure. There was no precedent to follow, and the parties did not agree on the rules for cancellation. The cancellation process for DDT clearly would be adversarial, with the pesticide industry already objecting.

Represented by Bill Butler, EDF insisted on judicial rules of evidence with qualification of expert witnesses, testimony to be relevant to the topics at issue, and full rights of cross-examination for all parties. That was a bottom line for EDF, and EPA lawyers agreed. If a witness was qualified as expert on topic A and not on topic B, he or she could testify on A but not on B. We had learned from experience that we did not want industry representatives and salesmen or lobbyists making opinion statements and then walking away, leaving a muddy record that would be little more than a popular vote open to varied interpretations. Industry wanted that. EDF wanted competent scientists to build an accurate record, and after months of haggling, EDF and EPA ultimately prevailed. There would be judicial rules of evidence. It was a triumph for Bill Butler and EDF. Little did we know then that this procedure would influence pesticide regulation by EPA far into the future. Judicial rules of evidence proved critically important in the litigation and eventual banning of aldrin, dieldrin, chlordane, heptachlor, and mirex, as we will describe in Chapter 12. Qualified scientists and experts testified in those proceedings, and some previously vocal advocates never appeared.

Since EPA had been ordered by the court to consider cancellation of all registrations of DDT, it was the DDT proponents who were bringing the appeal. They were known as the Group Petitioners. *It had become the burden of industry to prove DDT safe, whereas before it had been our burden to prove hazard.* This was not a minor matter. Twenty-seven pesticide manufacturers appealed the cancellation process, including the Montrose Chemical Corporation, the major manufacturer of DDT in Los Angeles. They were joined by the National Agricultural Chemicals Association and the USDA, all of them defending DDT and represented by their teams of lawyers.

Major precedents had again been established by the DDT litigation. Industry had the burden of establishing the safety of their pesticide products under judicial rules of evidence. Now, 40 years later, pesticides are still subject to more stringent regulation by EPA than the myriad of other nonpesticidal industrial chemicals. The DDT litigation set the standard that has been followed since 1971.

It was August 17, 1971, more than five years after our campaign against DDT began locally on eastern Long Island, and the hearing on the cancellation of DDT was about to begin before EPA. We anticipated that this proceeding should lead to a national decision. The hearing was to be held in a satellite office of the Interior Department's Bureau of Mines in Alexandria, Virginia. The hearing examiner was Edmund Sweeney, on loan from the Bureau of Mines, who seemed to know little about science and the environment and didn't seem to care much.

Sweeney was the only administrative law judge available at that time. The satellite office was chosen because it was within walking distance of Sweeney's home and he could walk home for lunch. The beginning of this hearing was not accompanied by as much fanfare, attendance, or publicity as had accompanied the onset of the hearing in Madison, Wisconsin, although there was a sizable cast of players.

On our side of the aisle defending cancellation of DDT was EPA with its lawyers and a couple of scientists. EPA was backing EDF on cancellation of DDT, whereas EDF wanted both cancellation and suspension. EDF was represented by Bill Butler and John Dienelt. Bill had never actually practiced trial law (his words), although he had clerked for a federal judge. Bill and John also represented EDF's partners, who joined the proceedings in support of EDF and helped to defray some of the administrative costs: the Sierra Club, the National Audubon Society, and the West Michigan Environmental Action Council.

John F. Dienelt had been hired by EDF during the summer of 1971. He had been a classmate of Bill Butler at Yale Law School, had clerked for a federal judge, and had come to EDF from the Solicitor General's Office of the U.S. Justice Department. (The Solicitor General's Office represents federal agencies before the Supreme Court.) John focused primarily on the human health aspects of the DDT case; Bill concentrated mainly on the environmental issues.

The EPA lawyers and scientists were the same people who had worked for the Pesticides Regulation Division when it was within USDA. We enjoyed having lunch with them and plotted strategy together, and they called as EPA witnesses a number of the scientists we recommended, which saved EDF travel expenses. Scientific support for Bill and John came primarily from Dr. Ian C. T. Nisbet of the Massachusetts Audubon Society, Dr. Robert W. Risebrough from the University of California at Berkeley, and me. All three scientists were not usually there at the same time. When in Washington I usually slept in the Butlers' basement with Schnoofie, Bill and Helga Butler's delightful, de-scented pet skunk.

Much of our strategy was similar, and some of the witnesses were the same as in the Wisconsin hearings. I will try to emphasize the differences, however, and not be overly repetitious. In the two years between the hearings much new data had appeared. We were leading the science as new developments were unfolding. The understanding of DDT's role in causing thin-shelled eggs had advanced considerably, and the potential carcinogenesis and mutagenesis of DDT had become an important issue in the case. Public opinion seemed more concerned with cancer than with peregrines.

The conduct of the hearings in Wisconsin, however, was much superior to the Washington proceedings because of the difference in the competence and temperament of the respective hearing examiners. Taking it in Bill Butler's words from *Acorn Days*:

> The hearings were a disaster. Sweeney could not understand what was happening, and about halfway through the hearings he informed me that I couldn't cross-examine the witnesses anymore because he already knew my views and didn't need to hear the questions again. This required EPA Assistant General Counsel Alan Kirk to appear and read a statement to Sweeney to the effect that this was not the way the Agency anticipated proceeding. Whereupon Sweeney sulked for several days trying to decide whether he was prepared to continue with the hearings or not. He subsequently did continue but in very bad temper.

Sweeney often seemed abusive of witnesses, demanding an inappropriate "yes or no" answer from scientists, where neither "yes" nor "no" was a correct answer. At one point EPA refused to allow its own scientists to appear in this hostile environment. Sweeney appeared to favor industry positions on issues. Following one contentious debate, Sweeney chastised Bill Butler: "You know, Mr. Butler, if you were six years old, I'd know how to handle you." John confirmed Bill's impressions of Sweeney, biased against every EDF/EPA witness from the start and heavily favoring industry witnesses. After all, we were talking about "just birds" (John Dienelt, personal communication, March 17, 2014).

The quality of the hearing was not helped by Sweeney's behavior. Although Sweeney didn't seem to put much stock in the science presented by witnesses, he did appear to like some witnesses more than others. He apparently took to Joe Hickey immediately as a fellow Irish Catholic and potential drinking buddy to swap stories with at day's end. After Professor Hickey, an ornithologist from the University of Wisconsin and an authority on falcons, had testified for several days about the decline of the peregrine, Sweeney said to him, *"You know, Doctor, there is one thing I don't understand. What would happen if these duck hawks [the former name for Peregrine Falcon] all became extinct? Who would care, since, after all, we all like ducks?"* That was the level on which Sweeney conducted the hearings and subsequently wrote his opinion.

The Group Petitioners representing the DDT industry presented their case first, beginning the hearing with their defense of DDT. To maintain current registrations and reverse the EPA cancellation order, they had to prove that DDT, when used according to labels, did not harm humans or other nontarget organisms. The thrust of the EDF case had been that DDT is an uncontrollable substance once out of the can and that there is no label that can be written to provide such protection. The industry argued that DDT had been used for decades without harm to any individuals and that evidence of harm to wildlife was very weak. DDT, they asserted, had made many contributions to human health by preventing diseases, and DDT was allegedly still needed for insect control in various agricultural crops.

Wherever possible in this book I have followed chronological order, but for this hearing I will organize the testimony by subject. That should be easier to follow, since witness appearances were often scattered in time and topics, resulting in somewhat of a jumble. I have therefore organized testimony by subjects as follows:

DDT: Global Contamination Problem
Predatory Birds Pay the Price
Mechanism for DDE-Induced Thin-Shelled Eggs
Dangers of DDT to Fish and People Who Eat Them
Superior Insect Control Without DDT
Does DDT Cause Cancer? In Mice? Rats? Humans?
DDT Industry Shortage of Qualified Witnesses

DDT: GLOBAL CONTAMINATION PROBLEM

Drs. George Woodwell, Bob Risebrough, and several other scientists testified about the mechanisms whereby DDT residues travel around the world in currents of air and water, then contaminate food webs and many kinds of organisms remote from the site where the DDT had been applied. Woodwell, an ecologist with Brookhaven National Laboratory, outlined how radioactive debris from atomic weapons tests in 1954 had followed such circulation patterns and then was picked up by small plants and passed up the line to small animals and then larger animals, including tuna and other important human food fish. The U.S. Atomic Energy Commission had studied these processes exhaustively in the South Pacific following the tests (Woodwell, 1967).

Pollutants can follow similar routes. DDT has a very low vapor pressure, but it is not zero, so DDT residues can slowly vaporize into air by sublimation and co-distillation (Wurster, 1972). It similarly has a very low solubility in water, but it is not zero; DDT is slowly picked up by water. Minute DDT particles can also be transported by air and water. Currents of air and water can transport DDT residues from their source to all parts of the globe. DDT falls with rain and snow, gets into food webs, and travels up food chains, causing the highest concentrations in carnivores at the top (Risebrough, 1968; Rudd, 1964; Woodwell, 1967).

Results of these processes had been demonstrated repeatedly in years before the hearings. DDD added to Clear Lake, California, either killed or led to reproductive failure in Western Grebes (Rudd, 1964). DDT applied to marshes on Long Island, New York, spread throughout the food chains and concentrated in top carnivores, including Ospreys (Woodwell, Wurster, & Isaacson, 1967). DDT sprayed onto cotton fields in the southeastern United States contaminated Bermuda Petrels feeding on surface organisms hundreds of miles at sea in the Western Atlantic (Wurster & Wingate, 1968).

DDT runoff into Lake Michigan endangered salmon reproduction in the lake and in hatcheries (Macek, 1968a). Many predatory birds became contaminated, with catastrophic results (see below). Organisms in the Earth's most remote regions became contaminated: Marine mammals in the Arctic and penguins in the Antarctic had been found to carry DDT residues (Sladen et al., 1966). And finally, most, if not all, humans worldwide carried low levels of DDT residues, mainly as the metabolite DDE (World Health Organization, 2011). In short, DDT was a hazardous global experiment with both known and unknown consequences.

George Woodwell described the subtle, unpredictable, and yet important ways in which pollutants such as DDT degrade ecosystems. He compared the influence of radiation with that of pollutants and other disturbances that change and simplify ecosystems, changing their species composition and structure. Jerry L. Mosser of the Marine Sciences Research Center, State University of New York at Stony Brook, described how DDT changed the species composition of a phytoplankton community, favoring green algae over diatoms (Mosser et al., 1972). He speculated that changing the base of food webs could alter the path of nutrients, potentially leading to jellyfish, for example, instead of edible fish at the top.

Although high-visibility top carnivores got most of the research attention and publicity, detrimental effects also were occurring at lower levels throughout food webs. Ecosystems lose biodiversity and become impoverished and simplified, becoming less rich (Woodwell, 1970; Woodwell et al., 1967). Pollutants, including DDT, tend to do less harm to generalists that survive on a wide variety of foods than to specialists, such as eagles and hawks. Reduced survival or reproduction among susceptible top carnivores, which receive the greatest concentration of DDT, tends to favor more resistant herbivores and scavengers, such as gulls, rats, pigeons, and cockroaches. Algae that survive around sewage and pollution do not turn into edible fish; low-quality fish and jellyfish are the more likely result. The complexity of such ecosystem changes are barely understood, but they are real and can be profound (Woodwell, 1970).

PREDATORY BIRDS PAY THE PRICE

Several eminent scientists detailed the precipitous declines of various birds of prey and seabirds. They included Drs. Joe Hickey, Bob Risebrough, and David B. Peakall, biochemist, and Tom J. Cade, ornithologist, both from Cornell University. They described how widespread the DDT disaster for predatory birds in North America and Europe had become (Cade et al., 1971; Hickey & Anderson, 1968; Ratcliffe, 1967; Risebrough, 1986). Many species were involved, all of them accumulating high DDT concentrations at the ends of long food chains because of biological concentration, the process first described for Western Grebes on Clear Lake, California, 15 years earlier (Carson, 1962; Rudd, 1964). The thin-shelled egg phenomenon caused by DDE, the DDT metabolite, was much better understood than it had been two years earlier in Madison. In many of the species studied and analyzed, DDE concentrations in the eggs were inversely correlated with eggshell thickness; as DDE concentrations went up, shell thickness went down (Risebrough, 1986).

It took a major scientific detective effort to understand what was happening, however, and there was not enough time and there were too few scientists to get all the answers. Imagine dozens of species, millions or billions of birds, spread out over two major continents, many in remote regions with few roads, and a relative handful of scientists trying to figure out what was happening, all within a few years. Some of these species, especially predatory birds such as the peregrine, have very thin and widely dispersed populations in wilderness areas, presenting a large logistical problem. It is remarkable that so much was learned about the complex DDT issue by a small number of dedicated and highly capable scientists, many of whom testified in these hearings.

Because of its special status among falconers, the Peregrine Falcon had probably received more attention by many ornithologists and other scientists, with more money spent on its recovery efforts, than any other of the many bird species affected by the DDT syndrome. Cade and Hickey summarized the historical record, outlining the dramatic decline into the early 1970s (Cade et al., 1988; Kiff, 1988). The peregrine was never an especially common species in North America, nor is it common in most of its worldwide range. Until the mid-1940s there were about 1,700 known nesting pairs in North America; the total at that time was estimated at 2,500 to 3,000 pairs.

From the late 1940s into the 1960s an unprecedented population crash occurred, less severe in some areas but to total extinction as a breeding species east of the Rocky Mountains in the United States. By 1975, only 324 known nesting pairs remained on the North American continent, about 19% of the original population (Kiff, 1988). Peregrines declined similarly in much of Europe (Risebrough, 1986). Peregrines are one of a few species of birds with a global distribution. Species extinction therefore did not threaten, although little was known about their status on continents other than Europe and North America.

Cade described analyses of peregrines breeding in Alaska that have different migration patterns and therefore different exposures to contaminants (Cade et al., 1971). The inverse correlation between DDE in their eggs and eggshell thickness was highly significant, and populations of the birds were declining. There was no question that the great decline of the Peregrine Falcon on two continents was caused primarily by DDE contamination leading to eggshell thinning. This decline was accelerated by direct mortality from dieldrin poisoning of adult birds, a factor considered especially important by some scientists (Cade et al., 1988; Nisbet, 1988; see also Chapter 12).

The story of one pair of peregrines nesting on Morro Rock, a 581-foot-high volcanic plug on the central California coast, is interesting and instructive. While just about all peregrines along the West Coast had been eliminated by DDT contamination by the late 1960s, this one pair continued to raise chicks, year after year. When Bob Risebrough climbed to inspect the nest, he found the remains of Mourning Doves, House Finches, and other seed-eating birds. This pair preferred to fly inland to feed on herbivorous birds, yielding a relatively short food chain and lower DDT contamination than peregrines feeding at sea on seabirds with longer food chains and greater DDT contamination (Risebrough, personal communication).

When the Bald Eagle became the national bird in 1782, there were as many as 100,000 nesting pairs in the lower 48 states. Their numbers had diminished to 417 pairs by 1963, to 487 pairs by another count in 1963, and probably to a lower number by about 1970, from shooting, lead poisoning, loss of habitat, hitting electrical wires, and DDT poisoning (Watts et al., 2008). Bald Eagles were essentially eliminated by DDT contamination from the Great Lakes region between the 1950s and the early 1970s (Best et al., 2010). Breeding success declined from 1.26 young per nest in Ontario to 0.46 young per nest in 1974. In Florida the number of active nests declined by 80% during the 1950s (Broley, 1958). Many authors documented the widespread decline of the Bald Eagle correlated with DDE contamination in the lower 48 states (Grier, 1982; Risebrough, 1986; Wiemeyer et al., 1984). Shell thinning and population decline of the closely related White-tailed Eagle in Scandinavia were similarly correlated with DDE concentrations (Wiemeyer et al., 1984).

The decline of the Osprey was believed caused mainly by DDT-induced eggshell thinning, which reduced the reproductive output of breeding pairs. The eminent ornithologist Roger Tory Peterson reported at the Madison conference in 1965 (Hickey, 1969) that the number of Osprey nests in southern Connecticut declined from 200 in 1938 to 13 in 1965, the latter fledging only two chicks. Rates of reproduction declined from a normal 2.3 young per pair per nest to 0.29. Eggs and dead chicks contained 16 to 100 ppm of DDT residues. Similar data on severely declining numbers of Ospreys and reproductive rates, with elevated DDT contamination, were reported from Long Island, New Jersey, the Great Lakes region, and the West Coast (Henny et al., 2010). The breeding population declined from an estimated 1,000 active nests in the 1940s between New York City and Boston to an estimated 150 nests in 1969.

Bob Risebrough described the near-total reproductive failure of the Brown Pelican (Fig. 10.1) colony on Anacapa Island in 1969, which resulted from the massive DDT contamination through the Los Angeles sewer system by the Montrose Chemical Corporation (Risebrough et al., 1971). The pelicans were feeding on highly DDT-contaminated anchovies. The nesting colony was littered with broken eggs, and shell fragments were 50% thinner than normal. A few unbroken eggs had only a dented membrane. Nearly 1,300 nests had produced a grand total of two chicks.

Brown Pelican reproduction in Southern California collapsed, and the bird had become scarce along much of the West Coast by 1970 (Schreiber & Risebrough, 1972). Most of the pelicans that continued to be seen in California were breeding successfully along the coast of Mexico. Double-crested Cormorants also suffered breeding failure in Southern California from thin-shelled eggs (Gress et al., 1973).

Brown Pelican populations in the southeastern United States also suffered catastrophic declines in the 1950s and 1960s. In the Gulf of Mexico

Figure 10.1 Brown Pelican. Photo by Marie H. Gladwish, published with her permission.

they disappeared entirely except for small populations in Texas and western Florida (Wilkinson et al., 1994). They were gone from Louisiana, where the Brown Pelican is the state bird. Pesticides, including endrin and DDT, were a major factor in this decline, coming from the Mississippi River drainage of the agricultural Midwest. Pelican populations along the Atlantic coastal states also declined, mainly due to DDT runoff from cotton plantations, where DDT was heavily used. The Brown Pelican joined the Endangered Species List in 1970.

Many other avian species were involved in the DDT-induced decline. Already mentioned was the Western Grebe failure to reproduce at Clear Lake, California. The Cooper's Hawk, which feeds on birds in forests, suffered a population decline in North America, and the Prairie Falcon exhibited shell thinning (Risebrough, 1986). Merlins, another falcon, were also affected, as was the very rare, then nearly extinct California Condor. A rare oceanic seabird that lives far offshore, the Bermuda Petrel, had extensive DDE contamination and low reproduction rates (Risebrough, 1986; Wurster & Wingate, 1968). One addled egg had an abnormally thin shell, but sample sizes were too small to reach conclusions.

Eggshell thinning, low reproduction rates, and population declines were also recorded in Europe, including the Peregrine Falcon, Osprey, Eurasian Sparrowhawk, Golden Eagle, Merlin, Eurasian Kestrel, Eurasian Buzzard, and White-tailed Eagle (Risebrough, 1986). Dieldrin used in sheep dips and as seed dressings caused direct mortality in some of these species, accelerating the population declines. Other avian families with symptoms of the avian DDE disease were loons, storm-petrels, boobies and gannets, cormorants, anhingas and darters, herons, ibises, storks, ducks and geese, vultures, gulls, auks, owls, and crows (Risebrough, 1986). Based on the family- or generic-specific responses to DDE contamination, Lloyd Kiff of The Peregrine Fund estimated that as many as 10% of the world's bird species may have suffered significant eggshell thinning during the years of heaviest DDT use (Kiff, personal communication).

The DDT industry could not produce any credible witnesses who could counter this DDE/thin-eggshell evidence linked to avian reproduction. Instead, they attacked our witnesses with cross-examination, attempting to undermine their research. The implication was that DDT had provided great benefits to humans and that these were "just birds," with plenty of other birds of sturdier species remaining.

Despite several thousand scientific references on this topic, adequate data for all these families and species did not and still do not exist. Furthermore, few data exist from continents other than North America and Europe. There were too few ornithologists and other scientists scattered around the world, with too little funding, facilities, and time to fully document the extent of this disease. The steep decline in many of these bird populations occurred within at most two decades, and its magnitude was not appreciated until it was almost over with the banning of DDT in 1972. We will never know the full extent of the avian DDE disease.

MECHANISM FOR DDE-INDUCED THIN-SHELLED EGGS

In the Washington hearing, enzyme induction in the liver leading to estrogen metabolism appeared a less likely explanation for the mechanism causing thin-shelled eggs than it had in Madison. Instead, inhibition of the enzyme calcium ATP-ase was now believed to be the mechanism. Before egg laying, estrogen secretion by the ovary causes the female bird to accumulate and store calcium carbonate in the hollow parts of her skeleton, mainly the femur. At egg-laying time the calcium is carried by the bloodstream to the oviduct, where Ca-ATP-ase transports the calcium from outside to inside the oviduct, where the eggshell is formed. If Ca-ATP-ase is inhibited by DDE, insufficient calcium is available inside the oviduct to form an adequate shell.

Although the function of the enzyme Ca-ATP-ase is the same among all bird species, the susceptibility of the enzyme to the DDE/eggshell-thinning process differs widely between species. Thus predatory birds such as peregrines, Bald Eagles, Ospreys, pelicans, and cormorants are very sensitive to DDE, whereas pheasants, quail, and chickens are insensitive. Each species apparently has a threshold level of shell thinning. Below that threshold, the birds can reproduce; above it, eggshells can break and reproduction fails.

Enzyme induction leading to estrogen breakdown remains a possible explanation for altered reproductive behavior in the birds (Risebrough, 1986), but this factor is far less important than the DDE/eggshell thinning in explaining the population declines of the birds.

William H. Stickel, husband of Lucille Stickel who had testified in Madison, and Robert G. Heath, also from the Patuxent Wildlife Research Center in Laurel, Maryland, detailed the advances in understanding the thin-eggshell effect as developed in experiments at Patuxent (Heath, Spann, & Kreitzer, 1969). They clearly showed that DDE was the active agent causing thin shells. DDT also caused thin shells when fed to the birds because DDT was first converted to DDE in the birds. Other environmental pollutants tested at environmental concentrations, including dieldrin and PCBs, did not cause eggshell thinning (Risebrough, 1986, review chapter with many earlier references). DDT proponents had tried and failed to pin shell thinning on PCBs in Madison, and here was further confirmation that DDE, not PCBs, was the causative agent for thin-shelled eggs.

Commercial "DDT" is actually a mixture of several chemical compounds, and one of them, o,p´-DDT can also act as an estrogen, as Welch had testified in Madison (Welch et al., 1969). Feminization of males as embryos by DDT might explain the shortage of male Western Gulls on the nesting grounds in Southern California (Fry & Toone, 1981). Lacking an adequate number of males, many females paired with each other and laid double clutches of mostly sterile eggs. Only a small number of such nests produced chicks. This report was an early recognition of the now widely recognized field of endocrine disrupters, where trace amounts of many pollutants display feminine hormonal activity. This phenomenon was detailed by Dr. Theo Colburn in her path-breaking book *Our Stolen Future* (Colburn et al., 1996).

Thin-shelled eggs caused by DDE beyond a certain threshold break prematurely in the nest and produce no chicks. The industry did not produce a qualified scientist who could refute this evidence. Their nearest attempt at "proving" that DDT was safe for birds came from observations of increased sightings at Hawk Mountain in Pennsylvania on Christmas bird counts. Cross-examination revealed that the increases resulted from an increase in the number of observers rather than the number of birds.

DANGERS OF DDT TO FISH AND PEOPLE WHO EAT THEM

The testimony of several scientists detailed the hazards of DDT to fish and certain crustaceans (see Wurster, 1972, for references). As in his testimony in Madison, Macek told how DDT lowers the reproductive success of fish by accumulating in the egg yolk and killing the fry shortly after they hatch from contaminated eggs (Macek, 1968a). He further demonstrated the overwhelming significance of the food chain in the accumulation of DDT by trout (Macek & Korn, 1970). The occurrence of this effect in New York State, Lake Michigan, and the Gulf of Mexico was described by Macek, Philip Butler, Thomas Duke, Robert E. Reinert, and other scientists in this field. The DDT proponents were unable to dispel the hazards of DDT to fish. Witnesses on chemical carcinogenesis detailed the hazards to humans who eat fish contaminated with DDT, as described later.

SUPERIOR INSECT CONTROL WITHOUT DDT

The success or failure to control insects was the topic of much testimony. With the largest use of DDT being on cotton, the industry asserted that DDT was "essential" to control the cotton bollworm. Witnesses from both sides agreed, however, that the bollworm and several other secondary pests usually become pests only *after* insecticide treatments. Some insecticides, including DDT, kill the natural enemies of the bollworm; bollworm populations then increase to damaging levels (Huffaker, 1971). *The industry argument therefore was that DDT was needed to control a pest problem that was actually caused by DDT (or some other insecticide) in the first place.* In California, a major cotton-producing state, integrated control preserved the natural enemies of the bollworm, achieving increased cotton yields without any DDT (van den Bosch, in Huffaker, 1971).

"Nightmarish" insect problems caused by DDT and other insecticides in cotton crops in the Canete Valley, Peru, were described by Bob van den Bosch ("Van") (Smith & van den Bosch, 1967). DDT applications began in 1949, and by 1955 the natural enemies of the pest species had been destroyed, new species rose to pest status, serious resistance to DDT had developed, and crop yields declined. More frequent applications of more toxic materials only made matters worse. These severe conditions came on especially rapidly because the Canete Valley is an isolated ecosystem surrounded by a rainless desert with no vegetation and few nonresistant insects or natural enemies to repopulate the cotton fields.

Desperate farmers appealed to their experimental station, which advised major changes leading to record-high cotton yields by 1957–1963. They eliminated the use of DDT, cut milder insecticide treatments back to 25% of the dosages recommended by the manufacturer, reintroduced natural enemies of the pests from other valleys and allowed them to reestablish, and made other cultural changes to restore the ecological balance in the valley. Van called the Canete Valley experience a classic case of excessive reliance on broad-spectrum insecticides leading to severe insect pest problems, increased costs, and falling crop yields. The unique circumstances of this valley served as a predictor of things to come elsewhere.

Use of DDT and other broad-spectrum insecticides is incompatible with integrated control because they destroy natural enemies, inducing outbreaks of secondary pests that are often worse than the target pest. Farmers, consumers, and the environment all benefit from integrated pest management (IPM), a high and stable crop yield is achieved, and pesticide sales decline. One would think that all cotton in California would be under IPM, but most is not. Van told how the pesticide industry stirs up animosity toward IPM among growers by spreading disinformation, confusion, and false stories, along with weekend fairs with free food and entertainment. Excessive pesticide use, high costs, increased pest problems, reduced crop yields, and lower profits for growers are the results, but industry pressures keep growers in the dark and acting against their own best interests.

Van explained how pesticide manufacturers fund university research on pesticides and insect control leading to recommendations for the use of their own pesticide products. They manage the pest control advisory system to their financial advantage through political connections with USDA. Van described all this and the conflicts of interest later in his book *The Pesticide Conspiracy*, published only a month before he died in 1978 while jogging in Berkeley (van den Bosch, 1978). Clearly the pesticide industry and its (previous) regulatory agency USDA were objecting to IPM partly because it would sell less pesticide. Fast-forward 40 years: Is there a parallel here with today's fossil fuel industries that oppose greater energy efficiency in cars, buildings, and other energy-consuming systems that would benefit consumers and the environment and diminish climate change, but would sell less oil and coal? History repeats itself!

In 2012 I wanted to ask how IPM had progressed since the DDT ban. Three outstanding scientists—Don Chant, Paul DeBach, and Van—had testified 40 years ago in Wisconsin and Washington that it was a better way to control insect pests, and no DDT was a part of it. But alas, these splendid people are deceased. Dr. Andrew P. Gutierrez, Professor Emeritus of Ecosystem Science at the University of California at Berkeley, was a graduate student, postdoc, and close academic colleague with Van. (Andy and Van had a brief conversation the day Van died. Andy thought he looked thin, and he had a slight cough. "Van, you shouldn't go jogging until you feel better," said Andy. "I'm in great shape," said Van. He had seen his doctor only the day before. There was a twinkle in his eye as he jogged away. Andy never saw him alive again.)

Here was Andy's response to my questions about the next 40 years:

The scientific and legal effort to ban DDT and related chlorinated hydrocarbon insecticides was a watershed period in pest control because it exposed not only the ecological and health dangers associated with their use, but also questioned the basis of the purely chemical control strategy in pest control, as outlined by van den

Bosch (1978). It became the legal precedent for challenging other chemical pesticides. The integration of chemical, cultural and biological control to minimize environmental impacts and maximize yields and profits came to be known as integrated pest management (IPM), which was widely adopted by most scientists and universities. Van called it a better way to battle bugs. Pest control is better now with less insecticide because earlier problems were caused by insecticide overuse and misuse (Huffaker & Gutierrez, 1999).

DOES DDT CAUSE CANCER? IN MICE? RATS? HUMANS?

Emphasis on carcinogenesis and mutagenesis by DDT was much greater in the Washington hearing than it had been in Madison. The industry correctly claimed that DDT had been important in fighting malaria and typhus in the past, but they could not find any evidence, or find a witness who would say, that DDT was still needed for disease control in the United States. Wayland Hayes and his colleague, Edward Laws from Johns Hopkins University, described their several-year studies on men heavily exposed to DDT, as Hayes had in Madison. He again asserted that DDT is "absolutely safe" for humans, a claim directly at odds with other expert testimony (Hayes, Durham, & Cueto, 1956; Laws, Curley, & Biros, 1967).

Cross-examination again showed that only men were involved in these tests, there were too few men for statistical significance, the time frame was too short, and there was no systematic search for tumors. These were not carcinogenesis tests. They were tests showing only that DDT was not acutely toxic to men. These tests did not dispute the assertion that DDT was a possible human carcinogen, based on animal experiments, the only accurate assessment of carcinogenesis we have.

The DDT proponents repeatedly argued that only test results with human subjects could determine the safety of a chemical and that DDT had passed such tests. But several witnesses made it clear that valid human tests could not be performed and that proof that DDT was or was not a human carcinogen could not be demanded. *Testing for carcinogenesis using human subjects is impractical, insensitive, and immoral.* It would require tens of thousands of volunteers for statistical significance and would take decades, and the results would be confounded by exposures to numerous other substances. Human cancers are diagnosed every day, and most appear with no known cause. The question of what initiated the tumor is rarely asked and almost never answered. Only specific cancers involving very large numbers of people can be traced to their cause, such as smoking, asbestos or diethylstilbestrol exposure, and sun damage, but by then it is too late.

A joint USDA/EPA witness was Lorenzo Tomatis, an official (and soon to be director) of the International Agency for Research on Cancer in Lyon, France. Some of the most definitive experiments with DDT had been done in that laboratory, and Tomatis was a world-class authority on chemical carcinogenesis. Tomatis stated that *DDT was definitely carcinogenic in several strains of mice, and in several laboratories, making DDT clearly a human health hazard* (Innes et al., 1969; Tomatis, Turusov, Day, & Charles, 1972).

Top authorities in carcinogenesis, including Tomatis, Umberto Saffiotti from the National Cancer Institute, and Samuel S. Epstein (an EDF witness) from Case Western Reserve University Medical School, described and interpreted the standard test for carcinogenesis. Pure strains of mice or rats are substitutes for people, and using very high, maximally tolerated dosages of the material to be tested greatly increases the sensitivity of the test, allowing a manageable number of animals to achieve statistical significance. The test takes about two years. The animals are then sacrificed, dissected, and thoroughly searched for tumors. This is the standard test for evaluating potential carcinogens and is very accurate. Can you imagine running such tests on human beings?

There is much public confusion about these tests, sometimes fostered by claims made by the manufacturers of the chemicals. Contrary to such claims, *high dosages do not turn noncarcinogens into carcinogens; "everything" does not cause cancer at high doses. Carcinogenic chemicals are relatively rare.* Almost all known human carcinogens give a positive test with rodents in such experiments. All materials that give a positive

rodent test, however, cannot be claimed with certainty to be human carcinogens because we cannot run the human test. We can only conclude
that substances that cause tumors in laboratory animals are possible or
probable human carcinogens. To be safe, such materials should be kept
out of the human environment. DDT was such a high-risk chemical, said
witnesses.

Most human cancers are environmentally induced in genetically sensitive individuals, usually not appearing for years or even decades after exposure to cancer-causing agents (carcinogens) found in air, water, food, or
workplace (Epstein, 1970, 1972; Saffiotti et al., 1970). Rates of cancer causation may be very low. If one person in 10,000 exposed individuals were to
get cancer, the risk might sound trivial, yet if all Americans were exposed to
such a potential carcinogen, then 20,000 cancers (31,000 in a 2012 U.S. population) would result, a public health catastrophe. *It is critical, then, to identify
such a "needle in a haystack" and to keep it out of the human environment.*

The Group Petitioners for DDT maintained that mice are not relevant
for evaluating human hazards; that there is a safe threshold for carcinogens; that "everything" causes tumors at high doses; that the mouse
tumors were benign rather than malignant; and that there was no proof
that DDT causes cancer in humans, along with many other allegations.
Competent witnesses in chemical carcinogenesis rejected every single one
of these contentions. *There is no basis for establishing a "safe threshold"
for carcinogens,* a recognition embodied in the zero tolerance for food additives in the 1958 Delaney Amendment to the Federal Food, Drug, and
Cosmetic Act, the law then in force.

The original EDF petition to the U.S. Department of Health, Education, and Welfare (HEW) did not seek an immediate enforcement of a
zero tolerance for DDT in foods, since that could not be achieved without
barring large segments of our food supply. The petition spent 40 pages
asserting that the Delaney clause firmly established that no additive that
caused cancer in laboratory animals can safely be added in any amount
to the food supply.

The petition also outlined the substantial legislative history affirm-
ing that the Delaney clause applies to unintentional food additives,

including DDT. DDT gets into foods because it is uncontrollable, once out of the can. EDF's petition therefore demanded that the Secretary of HEW "take all action that is necessary to effect the elimination of . . . [DDT] . . . from raw agricultural commodities." Translated, that did not mean confiscating our foods; it meant not putting any more DDT into our foods. The ultimate Ruckelshaus decision accomplished exactly that objective. The carcinogenesis of DDT also placed it in violation of the protective clauses of FIFRA, which eventually formed the basis of the Ruckelshaus decision.

The experiments on men described by Wayland Hayes suggested a low acute toxicity of DDT for men but have no relevance for evaluating its cancer-causing potential. Competent testimony that DDT is a cancer hazard for man, as indicated by animal testing, was not successfully challenged by the industry.

Testimony by Dr. Marvin S. Legator from the U.S. Food and Drug Administration indicated that DDT also represents a potential genetic hazard for man. He described experiments in which dominant lethal mutations were increased in rats after the males had been exposed to DDT (Palmer, Green, & Legator, 1973). This mutagenesis hazard could not be dispelled by the DDT proponents.

DDT INDUSTRY SHORTAGE OF QUALIFIED WITNESSES

The DDT proponents had difficulty producing witnesses who could refute the testimonies of expert EPA and EDF witnesses. They called Dr. Thomas H. Jukes to the witness stand. Jukes had worked for American Cyanamid and then the University of California at Berkeley, mainly as a nutritionist. He had written numerous opinion letters to editors and articles extolling the virtues of DDT and opposing restrictions on it but had not actually worked with DDT or done any relevant research. Industry lawyers tried repeatedly to qualify him as an expert, but each time a substantive question about DDT was asked, Bill Butler objected and Sweeney sustained the objection (surprisingly). After a conference among all lawyers and

Sweeney, and despite Jukes's alleged offer to eat a pinch of DDT to "prove its safety," it was decided that Jukes was not a qualified expert on anything related to DDT, and he was sent back to Berkeley without giving any substantive testimony.

Another questionable witness for the DDT proponents was Dr. Norman Borlaug, winner of the Nobel Peace Prize as the developer of the Green Revolution in high-yielding crops. He was a plant geneticist and breeder. The crops he developed were especially susceptible to insect attack, so Borlaug became a promoter of insecticides, especially DDT. He considered EDF a "powerful lobby group of hysterical environmentalists." But he was not an expert on DDT, birds, environmental toxicology, cancer, insect control, or any other topic relevant to the DDT hearing. The DDT proponents had apparently invited him as a witness for public relations reasons, and his testimony dealt with the work he had done in foreign countries, not the United States.

When Borlaug was asked substantive questions about DDT, John Dienelt objected, but this time Sweeney overruled him: Apparently Sweeney wanted to hear what a Nobel laureate would have to say. The result was that Borlaug was permitted to testify about DDT, birds, wildlife, insect control, and other topics for which he was not qualified. Borlaug said that DDT was no problem for birds. Then John had to cross-examine him on these topics to weaken their importance in the transcript. After the hearing session, Borlaug covered a wide range of topics at a press conference outside the hearing room arranged by the Montrose Chemical Corporation, the world's largest DDT manufacturer.

After his press conference, Borlaug apparently headed for the White House. He had become friends with President Nixon, who greatly admired him. It was many years later that I learned more about the circumstances of Borlaug's presumed visit to the White House following his testimony.

I first met William Ruckelshaus at an EDF board meeting on Orcas Island in the San Juan Islands in Washington State in 1997, 25 years after his DDT decision. In connection with developing this book, he and I had lunch together on March 7, 2012. We discussed several matters, including the following story, which he recounted in an e-mail to me on March 12. Here is the bulk of that e-mail.

Dear Charlie,

Here is what happened regarding the possible interference by the President with my obligation to make the decision on DDT. I received a call from John Ehrlichman before the decision was made and before I had decided what to do. In this call, Ehrlichman recited the President's admiration for Norman Borlaug and indicated that Borlaug has expressed to the President some concern about what the decision on DDT would be. I believe it was based on that inquiry by Borlaug that the President expressed to Ehrlichman support for Borlaug and his concern about the final decision. Ehrlichman passed on that concern to me without telling me what the President wanted me to do or even suggesting the President was going to contact me about the decision.

Nevertheless, I was concerned that the President might try to direct the decision and I felt that was an interference with my obligation under the statute to decide what should be done with the pesticide. I believed this decision should be made without any political or other interference. As a result of my concern, I contacted John Mitchell whom I knew well at the Justice Department having served under him as Assistant Attorney General. Mitchell was then the Chairman of the Committee to Re-elect the President. We met in Lafayette Park, across the street from the White House and I told him of Ehrlichman's expression of the President's concern and made it very clear to him that I felt it was my decision, that I had not decided what decision was appropriate and that I did not feel it was a good idea for the President to get involved in this decision in any way. John Mitchell assured me he would pass that on to the President and I never heard another word from the White House.

Charlie, this is what happened and I felt that my discussion with Mitchell led to the White House backing away from any effort to interfere with the decision. Of course I don't know that they would have tried to interfere. I never talked to Mitchell again about the subject so I don't know from him whether he spoke to the President or not. That's what happened and if you want to put that in the book, it is fine with me.

I don't know whether Borlaug had a discussion with the President prior to the President's expression of concern to Ehrlichman about what the final decision would be. It may well be that the time frame you have outlined in the paragraph that you supplied to me is correct and that Borlaug talked to the President after he had testified and expressed his concern about the ultimate decision.

President Nixon was clearly responding to public pressure when he put together his environmental messages and suggested legislation to the Congress. While all this resulted in quite a record for Nixon, I don't personally believe it was a result of his deep interest in the subject but rather a response to the pressure from public opinion. In any event, his record was quite good regardless of the motivation and he deserves credit for that.

All the best, Bill

Borlaug, like Jukes, should not have been permitted by Sweeney to testify where only expert testimony on the subject at hand was properly admissible. Expert testimony must pass the tests of relevance and competence. Other would-be witnesses who never appeared, either in Wisconsin or Washington, included J. Gordon Edwards, a professor of entomology at San Jose State University, and S. Fred Singer, a professor emeritus at University of Virginia, neither of whom had ever done work relevant to the DDT hearings or circumstances. Both were persistent DDT defenders. Edwards was described as eating a teaspoon of DDT per week, disarming his audiences and thereby "proving" DDT to be safe. Nevertheless, they both held press conferences and were regular authors in such industry-oriented groups as the American Council on Science and Health (www. acsh.org), which calls itself a "sound science group," and 21st Century Science and Technology (www.21stcenturysciencetech.com), which publishes Edwards's papers, which have not been peer reviewed (Edwards, 2004).

After the Ruckelshaus decision banning DDT in June 1972, J. Gordon Edwards issued the following press release, which was then entered into the *Congressional Record* by Senator Barry Goldwater:

The Infamous Ruckelshaus DDT Decision [was] an abject capitulation to professional environmental extremists and a tremendous defeat for science and mankind.

The hearing lasted for eight months, consumed 9,312 pages, involved more than 125 witnesses, produced 365 documents and exhibits, and ended in March 1972. On April 25, 1972, the hearing examiner, Edmund Sweeney, stated that, in his opinion, the benefits of DDT outweigh its risk to human health and the environment. It was not carcinogenic and posed no danger to birds. No surprise there! But the hearing examiner can only recommend; he has no decision-making authority. That rests with the EPA administrator, William Ruckelshaus, and we were optimistic he would not accept Sweeney's recommendation and would rule otherwise.

11

Ruckelshaus Decides

All parties to the hearings knew that on June 14, 1972, at exactly 10 a.m., the door of the EPA administrator's office would open and out would come someone to distribute copies of the decision on the future of DDT. Nobody knew what was in it, but all parties figured there would be something they would not like and would therefore want to appeal it to an appeals court. Appeals could be heard by any of several federal appellate courts around the country. More important, the first appeal made to any court would likely determine the location or venue where the appeal would be heard.

The DDT proponents knew they had done poorly in the DC Court of Appeals, so they wanted to get their appeal out of DC; surely the cotton belt would be best. So they were waiting for that door to open with an open telephone line to the 5th Circuit Federal Court of Appeals in New Orleans, Louisiana. We knew what they were up to, so we were determined to file our appeal very quickly with the US Court of Appeals for DC, where we had done very well.

That was not a simple procedure. Cell phones did not exist in those days. The EPA administrator's door opened, the papers came out, and both appeals were rushed to the respective courts of appeal. Not a second was wasted to see what was in the decision. EDF attorney Bill Butler flashed the appeal on a pay phone, which had an open line to another pay phone in the DC Court of Appeals building near the clerk's office, where EDF secretary Marie Bauman filed the EDF appeal.

Each side claimed it had gotten to its preferred appeals court first. The DDT proponents said the case would move to New Orleans for the appeal. Much controversy and confusion ensued. Finally, it was decided that the clocks were not properly synchronized and that EDF had won the rapid communication derby: The case would stay in Washington, DC. We will never know how the case might have come out had it been heard in the cotton belt. As with all stages of the DDT wars, nothing ever came easily, and we soon learned that it wasn't over yet—not quite.

THE RUCKELSHAUS DECISION: DDT GETS CANCELED! ALMOST!

The Ruckelshaus "Opinion and Order issued June 14, 1972, concerning the registrations of products containing the insecticide DDT" was finally published in the *Federal Register* on July 7, 1972, in Vol. 37, No. 131, pages 13,369–13,376, more than seven large pages of fine print. Rather than paraphrase, I will quote some relevant sections directly from the opinion and order.

> This hearing represents the culmination of approximately 3 years of intensive administrative inquiry into the uses of DDT. . . . *I am persuaded . . . that the long-range risks of continued use of DDT for use on cotton and most other crops is unacceptable and outweighs any benefits.* Cancellation for all uses of DDT for crop production and nonhealth purposes is hereby reaffirmed and will become effective December 31, 1972, . . . except that certain uses for green peppers,

onions, and sweet potatoes in storage may continue on terms and conditions set forth in . . . this opinion and the accompanying order.

Peak use of DDT occurred at the end of the 1950s and present domestic use of DDT . . . has been estimated at 6,000 tons per year, . . . [of which] approximately 86 percent or 10,277,258 pounds of domestically used DDT is applied to cotton crops.

FIFRA . . . establishes a strict standard for the registration of pesticides. Any "economic poison" which cannot be used without injury to "man or other vertebrate animals, vegetation, and useful invertebrate animals" is "misbranded," and is therefore subject to cancellation.

I am convinced by a preponderance of the evidence that *once dispersed, DDT is an uncontrollable, durable chemical that persists in the aquatic and terrestrial environments.* Given its insolubility in water and its propensity to be stored in tissues, it collects in the food chain and is passed up to higher forms of aquatic and terrestrial life. There is ample evidence to show that . . . it will volatilize or move along with eroding soil. While the degree of transportability is unknown, evidence of record shows that it is occasionally found in remote areas or in ocean species, such as whales, far from any known area of application.

Laboratory tests have . . . produced tumorigenic effects on mice when DDT was fed to them at high levels. Most of the cancer research experts who testified at this hearing indicated that it was their opinion that the tumorigenic results of tests thus far conducted are an indicator of carcinogenicity and that *DDT should be considered a potential carcinogen.*

Mr. Ruckelshaus was unconvinced by the many arguments of the DDT proponents' witnesses that DDT was not a carcinogen. He specifically rejected "the increasingly familiar argument that exposure to any substance in sufficient quantities may cause cancer." The "everything is cancerous argument falls," said Ruckelshaus. Several DDT proponent witnesses, "while men of stature in their fields—toxicology and pathology—and knowledgeable about cancer treatment and diagnosis, are not specialists in cancer research as is Dr. Saffiotti." Umberto Saffiotti from the National Cancer Institute was a key EPA witness, along with other carcinogenesis experts, including Drs. Lorenzo Tomatis and Sam Epstein (EDF witness), all of whom testified similarly.

Ruckelshaus continued:

The evidence presented by . . . EDF compellingly demonstrates the adverse impact of DDT on fish and birdlife. Several witnesses . . . [reported] lethal or sub-acute effects on aquatic and avian life exposed in DDT-treated areas. Laboratory evidence is also impressively abundant to show the acute and chronic effects of DDT on avian animal species and suggest that DDT impairs their reproductive capabilities. The [DDT proponents'] assertion that there is no evidence of declining aquatic or avian populations, even if actually true, is an attempt at confession and avoidance. It does not refute the basic proposition that DDT causes damage to wildlife species.

I am persuaded that a preponderance of the evidence shows that *DDE causes thinning of eggshells in certain bird species.* The evidence presented included both laboratory data and observational data. Thus, results of feeding experiments were introduced to show that birds in the laboratory, when fed DDT, produced abnormally thin eggshells. In addition, researchers have also correlated thinning of shells by comparing the thickness of eggs found in nature with that of eggs taken from museums. The museum eggs show little thinning, whereas eggs taken from the wild after DDT use had become extensive reveal reduced thickness.

Benefits—I am convinced by the evidence that continued use of DDT is not necessary to insure an adequate supply of cotton at a reasonable cost. Only 38 percent of cotton-producing acreage is treated with DDT, although the approximately 10,277,258 pounds used in cotton production is a substantial volume of DDT and accounts for most of its use. The record contains testimony by witnesses called

by registrants and USDA attesting to the efficacy of organophosphate chemicals as substitutes for DDT and, long range, the viability
of pest management methods, such as the diapause program. . . .
There is evidence that organophosphates would not raise costs to the
farmer and might, indeed, be cheaper.

Ruckelshaus also considered the hearing examiner's opinion, especially because the DDT proponents in oral argument had claimed that
the examiner had the "opportunity to resolve contradictions in testimony
based on demeanor evidence." But Ruckelshaus disagreed: "Nowhere
does the Examiner state that his conclusions were based on credibility
choices. Whatever extra weight, then, that might be due findings based
expressly on a credibility judgment is not appropriate in the case before
me." Ruckelshaus gave little weight to Sweeney's opinion.

Ruckelshaus had this to say about the risk/benefit ratio of crop uses
of DDT:

The agency [EPA] and EDF have established that DDT is toxic to
nontarget insects and animals, persistent, mobile, and transferable
and that it builds up in the food chain. No label directions for use
can completely prevent these hazards. In short, they have established
at the very least the risk of the unknown. That risk is compounded
where, as is the case with DDT, man and animals tend to accumulate and store the chemical. These facts alone constitute risks that
are unjustified where apparently safer alternatives exist to achieve
the same benefit.

Accordingly, all crop uses of DDT are hereby canceled, except
for application to onions for control of cutworm, weevils on stored
sweet potatoes, and sweet peppers [on the Delmarva Peninsula].

There remains the question of the disposition on the registered
health and Government uses and other noncrop uses of DDT. It
should be emphasized that these hearings have never involved the
use of DDT by other nations in their health control programs.

EPA's final order by Ruckelshaus, dated June 14, 1972, was as follows:

> In accordance with the foregoing opinion, findings and conclusions of law, use of DDT on cotton, [many food crops], in commercial greenhouses, for moth-proofing and control of bats and rodents are hereby canceled as of December 31, 1972. Use of DDT for control of weevils on stored sweet potatoes, green peppers in the Del Marva Peninsula and cutworms on onions are canceled unless within 30 days users or registrants move to supplement the record. . . . Cancellation for uses of DDT by public health officials in disease control programs and by USDA and the military for health quarantine and use in prescription drugs is lifted.

The decision was carefully crafted to be consistent with the best available science and with applicable U.S. laws. Once released into the environment, DDT is an uncontrollable substance that contaminates natural food chains and gets into human foods as an unintended additive.

EDF and EPA witnesses had proved that DDT damages wildlife and that DDT was also shown to be carcinogenic in laboratory animals. DDT was therefore in violation of FIFRA, which provides that any "economic poison" that cannot be used without injury to "man or other vertebrate animals, vegetation, and useful invertebrate animals" is "misbranded" and is therefore subject to cancellation. The pesticide industry had no effective counter to this evidence. Cancellation of DDT registrations in agricultural crops was therefore justified and mandated.

Ruckelshaus took great care, however, not to cancel DDT for potential public health use in the United States. He was also very specific in not blocking the manufacture of DDT for export and in emphasizing that his decisions did not (and could not) apply to any other country. EDF in its appeal made no attempt to block the use of DDT for malaria anywhere.

In a recent note to me (July 9, 2013) following my inquiry, Mr. Ruckelshaus said, "To the best of my recollection, malaria was not mentioned in the EPA proceedings. . . . I have many times said I would have approved the use of DDT were I the EPA Administrator in a country struggling with malaria."

BOTH SIDES APPEAL THE RULING TO THE COURT
OF APPEALS

EDF appealed the exemptions from cancellation of DDT on sweet potatoes, green peppers, and onions. Within less than a year these registrations also were canceled. Alternative insect control procedures had been found. Cancellation of the use of DDT in agriculture was therefore complete. EDF's goals on DDT had been achieved. For public health purposes, EDF asserted that there were alternatives to DDT, but that if no effective alternatives could be found in connection with a public health emergency, DDT could be used.

The pesticide industry appealed the entire Ruckelshaus decision. Attorney John F. Dienelt represented EDF at the oral argument of this appeal before the court. *On December 14, 1973, the DC Court of Appeals rejected the industry appeal and upheld the Ruckelshaus decision to ban DDT in its entirety as "supported by substantial evidence."*

The DDT wars were over. The consequences would continue for decades. The driving force behind this major pesticide policy decision came not from elected or appointed public officials, but from a handful of volunteer private citizens.

PASSING THE TEST OF TIME

Not every member of Congress was pleased with the EPA ban on DDT, so in 1974 the Appropriations Committee of the House of Representatives requested a report from EPA justifying its actions. In July 1975 EPA issued a 304-page report, "DDT: A Review of Scientific and Economic Aspects of the Decision to Ban Its Use as a Pesticide." This excellent and comprehensive document concluded that all elements of the ban were fully justified and supported by scientific evidence, and if anything, the bases for the ban had actually increased (EPA-540/1–75–022).

On October 21, 1972, FIFRA was greatly strengthened by the Federal Environmental Pesticides Control Act, which gave EPA considerably more effective pesticide regulatory mechanisms than were previously available under FIFRA. Subsequent to this enhancement of FIFRA, two emergency uses of DDT were allowed by EPA and another was denied, in each case by a benefit/cost analysis as had been suggested by EDF. None of these involved a public health issue (EPA-540/1-75-022, July 1975).

In May 2001, delegates from more than 100 countries (including the United States, Canada, and all members of the European Union) signed the Stockholm Treaty on Persistent Organic Pollutants (POPs). In this international treaty, not only was DDT on the list of restricted or banned chemicals, but so were the other five pesticides that had been banned decades earlier in the United States as a result of EDF's actions. Their stories follow in Chapter 12.

Considering all the scientific and legal developments that have occurred since 1972, there is little doubt that if a decision were to be made today on DDT, it would be essentially identical to the ruling made by William Ruckelshaus 43 years ago. *The Ruckelshaus decision has withstood the test of time.*

THEY ARE AFTER US AGAIN—40 YEARS LATER. WHY?

The defense of dangerous products by their respective industries has become an unfortunately familiar pattern: Attack the scientists as "pseudoscientists," question their competence and motivation, and cast doubt on the credibility of the science itself ("junk science"), a pattern brilliantly described in *Merchants of Doubt* (Oreskes & Conway, 2010). Such attacks on scientists and science have involved essentially propaganda campaigns in such major public issues as smoking and health, asbestos, DDT and a few other pesticides, the artificial sweetener saccharin, lead in gasoline, chlorofluorocarbons and the ozone hole, and acid rain from fossil fuel burning.

In recent years we have experienced the mother of all propaganda campaigns by the oil and coal industries, denying that manmade climate change is occurring and opposing efforts to reduce global warming and climate change from greenhouse gas emissions, especially carbon dioxide, from the burning of oil and coal. Blocking funding for and downgrading all forms of clean energy research, development, and implementation has been an important part of the campaign. It was enough to turn the American public around, from favoring action against climate change to opposing it (Gelbspan, 1998; Pooley, 2010). *The DDT issue has been enlisted as a mechanism to discredit scientists and diminish their credibility and influence on the climate change debate, various regulations against pollution, and other environmental protection issues.*

One would think that a product (DDT) that disappeared from the marketplace 40 years ago without the predicted famine and pestilence would no longer be a topic of public discourse. Younger generations would not have heard of it, older generations would largely have forgotten about it, and there would be no public controversy. Yet such is not the case: The DDT issue has picked up steam during recent decades, propelled by numerous industry-funded think tanks and front groups.

In connection with the Madison DDT hearings, environmental scientists were charged with killing farm workers by advocating a switch from DDT to the highly toxic parathion. We never advocated parathion. Now in the 21st century we are accused of withholding DDT from malaria control and killing African children. Our victims apparently changed in the new century.

For example, the website of 21st Century Science and Technology (http://www.21stcenturysciencetech.com/) contains many articles about DDT. One by Marjorie Hecht refutes the many "Big Lies" about DDT. She claims that bird populations actually increased during maximum DDT use (they were "trash birds," mainly blackbirds) and that peregrines and Bald Eagles were threatened with extinction before DDT arrived. According to Hecht, eggshell thinning is unrelated to DDT and is caused by oil, lead, mercury, stress, dehydration, heat and cold, and humans interfering with nests. She denies that DDT has anything to do with cancer, citing studies of men who ate DDT for two years with no ill effects. In 50 years of usage, says Hecht, DDT has never been proven to harm human beings.

This and the other DDT articles on this website are filled with false and deceptive claims of the safety of DDT for wildlife and humans. Hecht goes further with the character assassination of environmental scientists, claiming they are "genocidalists" by banning DDT in developing countries, allowing malaria to kill millions of people.

Articles on this website by Hecht and others also label global warming a "hoax." Disinformation about DDT is being used to discredit scientists who favor actions to limit climate change by reducing fossil fuel consumption. I asked Hecht on the telephone where her organization gets its funding. Her reply: "Memberships. It's a labor of love."

Many other conservative front groups follow a similar line. The Heritage Foundation (www.heritage.org) also attacks environmental scientists for opposing the use of DDT for malaria in Africa, implying "racism" against black children in the DDT ban. "Since 1972's DDT ban, almost a billion people . . . have died due to preventable . . . malaria." "The DDT example is just one item of a laundry list full of environmental policies gone badly." "Tragically, the environmentalists' war against DDT has become a war against the world's poor." Heritage has an abundance of disinformation on DDT and has still more about "the Hoax of global warming," questioning "climate hysteria" while favoring increased oil drilling. Koch Industries, a major oil producer, is a large funder of Heritage and other similar groups (*New York Times*, October 6, 2013).

The Cato Institute (www.cato.org), founded by the Koch brothers (Brulle, 2013), uses the same arguments, calling the DDT ban "genocide" while suggesting climate change might be beneficial. "Mitigating the effects of global warming will be much cheaper than greenhouse gas emission reduction," claims Cato, referring also to "the web of regulations created by the EPA." According to the Competitive Enterprise Institute (cei.org), the Bald Eagle/DDT myth is "still flying high." The website speaks of "eco-freaks" and claims that Rush Limbaugh deserves the Nobel Peace Prize for correcting the misinformation on DDT. *To Junkscience.com, the DDT issue and global warming are both junk science.* Some of these campaigns were clever in their deception and successful in delaying action against their products, all to the detriment of society at large.

On Junkscience.com's website there is a story about DDT and malaria entitled "A Green Eco-Imperialist Legacy of Death," with a photo of William Ruckelshaus wearing a shirt inscribed "DDT, a Weapon of

Mass Survival" that you have to see to believe (http://junksciencearchive
.com/malaria_clock.html; still available as of Feb 14, 2015). According to
the story:

> Since Ruckelshaus arbitrarily and capriciously banned DDT, an es-
> timated 16 billion (a continuously running clock might be higher by
> the time you read it) cases of malaria have caused immense suffer-
> ing and poverty in the developing world. Of these largely avoidable
> cases, 106 million (another running clock) have died.
>
> Most of these deaths, the author continues, were children, fe-
> tuses, and pregnant women: "Infanticide on this scale appears with-
> out parallel in human history. . . . This is not ecology. This is not
> conservation. This is genocide."

Junkscience.com also has abundant material casting doubt on global
warming and climate change. There are many more websites and other
media outlets that follow this well-organized, coordinated attack on environmentalists, scientists, and science itself, using misinformation, disinformation, and total fallacies about DDT as vehicles to degrade the
credibility of environmental positions. The objectives are to cast doubt on
the climate change issue and all efforts to deal with it. The point of all this,
along with an attack on EPA's clean air regulations, is to maintain, if not
enhance, markets and profits for fossil fuel industries.

How are these front groups and their well-coordinated programs fi-
nanced? Follow the money. The Union of Concerned Scientists has written
a thoroughly documented 63-page report entitled "Smoke, Mirrors & Hot
Air; How ExxonMobil Uses Big Tobacco's Tactics to Manufacture Uncer-
tainty on Climate Science" (2007; http://www.ucsusa.org/assets/documents/
global_warming/exxon_report.pdf). Several of the front groups cited here
attacking the DDT issue have been funded by ExxonMobil, which distrib-
uted about $16 million to 43 front groups between 1998 and 2005.

According to a more recent analysis, from 2003 until 2007 the Koch
and ExxonMobil foundations were heavily involved in funding climate
change denial organizations, but since 2008 they no longer make publicly

traceable contributions. Overall, from 2003 to 2010, funding of climate denial pass-through foundations has risen dramatically, with 140 foundations funneling $558 million to almost 100 climate denial organizations. Koch and ExxonMobil funds are hidden within that mix. Their mission is to spread disinformation about climate change and oppose all efforts to reduce carbon dioxide emissions. Disinformation on the DDT ban is well dispersed throughout this material. Most of these groups are tax-exempt 501(c)(3) organizations (Brulle, 2013).

The objective in the current controversy about DDT is not to bring back DDT but to use false and fraudulent allegations about the DDT issue to discredit scientists' arguments against fossil fuel use and to weaken antipollution regulations by EPA. The losing false arguments to defend DDT years ago have returned without challenge on the Internet. DDT proponents were unable to refute evidence of wildlife damage and carcinogenesis 40 years ago, so they attempted to shift the issue to fighting malaria in developing countries. That was irrelevant to our case based on American law and domestic agricultural usage, but now those irrelevant arguments have returned unchallenged for a different objective.

Efforts by the industry to block our campaign to ban DDT decades ago were relatively mild compared with the aggressiveness of these more recent attacks. The DDT proponents initially considered us a public relations problem, realizing only much later that we had considerable depth in scientific support. We were "ahead of the curve" and caught them "with their pants down." We have to wonder: Would we have succeeded in this world of 2014? Our efforts 40 years ago coincided with the great environmental awakening of the 1960s and 1970s. The first Earth Day was in 1970. It appeared then to be the beginning of a great era of environmental consciousness and protection.

THE USE OF DDT FOR CONTROLLING MALARIA

The charge that environmentalists have allowed the killing of millions of Africans, particularly children, with malaria by withholding DDT from malaria control programs is a total fabrication. Let us return to the DDT wars of 40 years ago. Malaria was not an issue because malaria had largely been eliminated in the United States before DDT arrived. The DDT hearings and resulting decisions applied only to the United States, not to any foreign countries, and Ruckelshaus went out of his way to emphasize that: "It should be emphasized that these hearings have never involved the use of DDT by other nations in their health control programs," he said. Further, his cancellation order excluded public health uses, such as malaria, from cancellation in the United States, nor were exports of DDT barred in any way. EDF has supported use of DDT for indoor spraying in malaria programs overseas in areas where it is still effective and not rendered useless by widespread resistance.

Where does DDT fit into the control of malaria today? The World Health Organization (WHO) issued a report written by many international expert scientists, updated in March 2011, called "Strengthening Malaria Control While Reducing Reliance on DDT" (http://www.who.int/ipcs/capacity_building/ddt_statement/en/index.html). WHO promotes integrated vector management, combining locally adapted, cost-effective, and sustainable methods in the transition away from DDT. In some countries, DDT use has been reduced or eliminated. In others, indoor residual spraying of DDT remains indicated. The quantities used in such programs are generally minimal and below levels of concern for human health or environmental damage, and they are more than offset by the benefits of malaria prevention. Such quantities are far less than the wholesale broadcast of large amounts of DDT into the environment for eradication attempts and for numerous other purposes, including agriculture, as in earlier years. Such heavy use generates insecticide resistance, which compromises or destroys the effectiveness of malaria control programs.

Another WHO report (WHO/HTM/GMP/2011) states that in much of sub-Saharan Africa, indoor residual spraying with DDT and insecticide-treated bed nets serves as the most effective means for controlling malaria. DDT remains longer (6–12 months) on walls and ceilings than most other insecticides, and it also has repellant effects, sometimes driving mosquitoes out of the building. DDT is still needed for malaria control in high-transmission areas, especially in sub-Saharan Africa. EDF has consistently supported this limited use of DDT for malaria control where it is still effective.

In many areas, especially in West Africa, mosquito resistance to DDT is a serious problem, and it is aggravated when large amounts of DDT are used in agriculture. Resistance to DDT often resulted in the simultaneous resistance to other insecticides, such as pyrethroids (cross-resistance), so that when another insecticide was introduced, the mosquitoes were already resistant to that new pesticide. Widespread outdoor use of DDT in agriculture or for mosquito extermination has since been prohibited.

WHO also released (2011) a comprehensive 309-page report entitled "DDT in Indoor Residual Spraying: Human Health Aspects (Environmental Health Criteria 241)," prepared by teams of scientists worldwide, with numerous references to earlier work on the carcinogenicity of DDT. It concluded that DDE, the major metabolite in biological systems, is a probable human liver carcinogen, which would be consistent with many studies showing liver carcinogenicity of DDE in mice and rats. A 1991 report by the International Agency for Research on Cancer (Vol. 53, pp. 179–249) similarly concluded that DDT is a probable carcinogen. In specific malaria control operations, including indoor residual spraying, the increased health risk from DDT exposure to applicators and dwelling occupants is presumably outweighed by the effectiveness of DDT in malaria control.

To summarize, the DDT issue in the United States 40 years ago involved wildlife damage and human health risks in this country only. U.S. laws applied, and DDT usage in any other country was not at issue in these proceedings. The industry defending DDT tried to introduce foreign uses and considerations, but these were irrelevant and inapplicable, as Ruckelshaus carefully stated in his final decision. More recent blogs and websites have attempted to disregard EDF's consistent position on the issue of foreign use of DDT, even implying that EDF was withholding DDT from malaria programs, thereby killing African children. Such statements are false and deceptive and are similar to the propaganda tactics described in *Merchants of Doubt* (Oreskes & Conway, 2010).

12

Encores: Five More Bad Actors Were Dispatched

After DDT, EDF's next pesticide targets were aldrin and dieldrin, both made by Shell Chemical Company. We had sought to block a dieldrin application in Michigan in late 1967, shortly after EDF was founded. That action was partly successful, delaying the application for many months and resulting in the application of less dieldrin.

EDF was often accused of being against all pesticides, but that was never true. We were against DDT and several other persistent chlorinated hydrocarbons. Those pesticides were uniquely hazardous because they lasted for many years, traveled freely in the environment, and were ingested by animals and humans everywhere. They were very damaging to wildlife and—as we discovered along the way—they posed cancer hazards to humans. We were also against a purely chemical approach to pest control because integrated control techniques were more effective in controlling pests and contaminated the environment with less chemicals. We had a short list of pesticides that we identified as "bad actors"—all of

which were persistent chlorinated hydrocarbons—and we eventually succeeded in getting all of them banned, although this took several years and
an immense effort by our attorneys and scientists.

On October 16, 1970, EDF filed a legal petition with HEW requesting the establishment of zero-tolerance levels for aldrin and dieldrin in
human foods. The petition was written by attorney Edward Berlin and
included a comprehensive review affidavit by me with numerous scientific references (Wurster, 1971). A few weeks later, pesticide regulation
was transferred from HEW to the new EPA, and from then onward the
action was pursued through EPA. I was not present during the hearing
that followed. Instead Dr. Ian C. T. Nisbet, then the director of science for
the Massachusetts Audubon Society, provided scientific support and attended the hearing. He wrote the next two sections (on aldrin and dieldrin
and on heptachlor and chlordane), in which he describes what followed.

ALDRIN AND DIELDRIN ON TRIAL: SUSPENSION MID-STREAM

By Ian C. T. Nisbet

On March 18, 1971, EPA issued notices of cancellation for all registrations of aldrin and dieldrin (A/D), but the marathon hearings did not
begin until July 1973. As with DDT, EPA declined to suspend registrations of A/D on the grounds that the hazards were not "imminent," so
that uses of both pesticides would continue unchecked until a final decision on cancellation was reached. Although EDF contested EPA's decision not to suspend, EDF and EPA were in agreement on cancellation and
worked closely together throughout the hearings. William A. Butler and
Jacqueline M. Warren were the attorneys for EDF. I worked part-time as
scientific advisor to EDF, alternating with my regular job at the Massachusetts Audubon Society. I stayed at the Butlers' house on Capitol Hill
with Schnoofie the skunk and commuted to EPA headquarters by bicycle.

Aldrin and dieldrin were always considered together because they are
closely related chemically and aldrin is rapidly converted to dieldrin in

the environment. Hence, all residues in the environment and all the toxic effects on wildlife were attributable to dieldrin, even where aldrin had been applied. Like DDT, A/D caused massive adverse effects on wildlife when they were widely used in the 1950s and early 1960s. Also like DDT, pest insects quickly developed resistance to them. Most uses were phased out during the 1960s and by 1973 only one major use remained—aldrin on cornfields in the Midwest to control soil insects. Usage in the United States had declined to 10 million pounds per year in 1973 from 22 million pounds in 1966 (Nisbet, 1988).

Cancellation hearings started in Washington in August 1973 and continued for more than a year. An experienced administrative law judge, Herbert L. Perlman, understood science and ran the proceedings in a professional and impartial way. The manufacturer, Shell, was well prepared and presented witnesses who were well qualified and often experts in their fields. One of the senior attorneys from the law firm representing Shell had even sat through most of the DDT hearings to learn how EDF and EPA presented their cases.

Shell's parent company was headquartered in the Netherlands, their main research laboratory was in the United Kingdom, and many of their expert witnesses came from those countries. A/D had caused massive wildlife damage in Europe in the early 1960s and Shell had long experience confronting critics. They had strategically withdrawn the most damaging uses, and their spokesmen had long experience in arguing away the scientific evidence for wildlife damage from the remaining uses; they brought these skills to the U.S. hearings.

For years Shell had cultivated a "green" image, promoting their scallop shell logo, supporting the publication of a major bird book, and disarming their opposition by discreetly funding major ornithological and conservation organizations in the United Kingdom. Besides their own research laboratory, they had strategically supported scientific research at universities and field stations, gaining the loyalty of experts in disciplines such as entomology, soil chemistry, pesticide metabolism, and bird populations. However, Shell had not previously had to subject its products to rigorous judicial scrutiny, nor to disclose the raw data from its research laboratory. Shell's scientific witnesses were surprised by the intensity and detail of the cross-examination that they encountered. The judicial rules of evidence as negotiated by EDF's Bill Butler for the DDT hearings were carried over with great effectiveness to the A/D hearings and the several other pesticide hearings that followed.

Detailed testimony presented at the hearings covered the chemistry of A/D, their breakdown products, patterns of use, efficacy in controlling pests, environmental transport and fate, levels of contamination in the environment, effects on wildlife, and human exposure via residues in food. Like DDT, A/D were persistent, mobile, and retained in animal tissues; residues of dieldrin were found everywhere, including wildlife far from points of use and the tissues of almost the entire human population of the United States (Wurster, 1971), if not the world.

Like DDT, A/D had caused massive wildlife damage in most areas where they had been used, as well as in many off-site locations. Unlike DDE, the major toxic effect of dieldrin in mammals and birds was lethal toxicity rather than effects on reproduction. Residues of dieldrin were stored in the fat of mammals and birds, building up over long periods of exposure until the animal encountered a period of stress and used up its stored fat, releasing the dieldrin into the bloodstream, where it was transported to the brain, causing convulsions and death. This pattern had been observed in many wildlife populations in Europe and North America and had been duplicated in an experiment with American Kestrels by Stanley Wiemeyer of the Patuxent Wildlife Research Center in Maryland (Wiemeyer et al., 1986). Birds of prey and fish-eating birds often died from dieldrin poisoning at points far removed from the places and times where they were exposed.

In Europe, the most damaging use of A/D had been as seed dressings on wheat, which caused massive mortality of pigeons, pheasants, and small mammals, as well as birds of prey and scavengers. The second most damaging use was in sheep dips, which killed fish downstream and many birds that fed on sheep carcasses, including Golden Eagles and the iconic Red Kite, which was almost extirpated from its last stronghold in the United Kingdom. Neither of these uses was important in the United

States. Here the most damaging uses were as seed dressings on rice, which killed many ducks, geese, and ibises in Louisiana and Oklahoma; for mothproofing, which killed fish and fish-eating birds downstream from woolen mills; and in the futile attempts to eradicate fire ants and white-fringed beetles in the 1950s, which killed almost everything, as described in *Silent Spring* (Carson, 1962).

Shell could have chosen to acknowledge these effects, arguing that those events were in the past and that uses at issue in the hearing posed minimal or no risks. Instead, they chose to defend almost all former uses, arguing that dieldrin was rapidly degraded in the environment, that residues of dieldrin in the tissues of the poisoned wildlife were not high enough to have killed them, or that other chemicals might have been involved. None of these arguments survived cross-examination or the weight of contrary evidence.

It proved difficult for EPA and EDF, however, to demonstrate that the main use of A/D on cornfields in the Midwest had led to serious wild-life damage. Few direct studies had been done linking those uses in the Midwest to effects, largely because there were few wildlife left to study in the areas of application or the downstream waters. The best-documented effect was on Great Blue Herons along the Mississippi River. Bald Eagles, egrets, and other fish-eating birds had also been killed. Powerful testimony by Dr. Robert L. Metcalf of the University of Illinois described his elegant experiments documenting the environmental fate and effects of dieldrin in laboratory "microcosms" that simulated natural ecosystems. He also described the widespread contamination of the Midwest environment with residues of dieldrin and its effects on native wildlife. This was courageous testimony; at that time the pesticide industry had great power within Land Grant universities and faculty members critical of pesticides often suffered retaliation.

By July 1974, more than 200 witnesses had testified in the hearings, with thousands of exhibits and 35,000 pages of transcripts. The evidence about environmental contamination and ecological effects was complete, and all parties had analyzed the record and submitted voluminous briefs. Because of the complexity of the record and the rather limited evidence connecting adverse effects to the specific uses that were in contention, it was not clear at that time whether EDF and EPA would have prevailed on the wildlife issues alone.

That question was never tested, however, because soon after the human health case started, *EPA abruptly suspended all uses of A/D*. The EPA administrator's opinion justifying this decision cited an "imminent hazard" to human health, based on new evidence that dieldrin caused cancer in animals and that residues of dieldrin had been found in 98% of the human population of the United States. This evidence was not truly "new": A study from Shell's own laboratory published in 1972 had reported that dieldrin induced liver tumors in mice, and residues in human fat had been reported since the early 1960s. The new evidence was corroborative studies in mice and rats and a systematic, large-scale study of residues in human tissues (Epstein, 1975).

This was the first time that EPA had suspended a pesticide, and some of the rules had to be formulated on the spot. FIFRA required a hearing within 15 days, but in fact, the hearing lasted for four weeks in August and September 1974. It involved intense activity: Dozens of witnesses on each side were prepared, presented, and cross-examined; thousands of pages of exhibits were introduced into evidence; and the transcript filled more than 10,000 pages. All that material had to be reviewed and summarized, with briefs to be submitted within a week after the hearing closed. With its greater resources, EPA shouldered much of that burden, and EDF contributed mainly by preparing witnesses and cross-examining. During the suspension hearing, I acted as the principal scientific coordinator for EPA.

The first legal issue to be argued was the meaning of "imminent hazard." Shell argued that an increased incidence of cancer 20 or more years in the future could not be "imminent." EPA argued that "imminent" did not have to mean immediate; the issue was whether current uses were increasing the hazard. This view eventually prevailed.

The scientific and legal issues in the suspension hearing were quickly narrowed to "corn and cancer"—whether the use of aldrin in cornfields led to significant human exposure to dieldrin, and whether that exposure significantly increased the risk of cancer. The connections became clearer as the hearing progressed: The use of aldrin led to high residues of dieldrin in soil. Although little dieldrin was taken up by corn, it was readily taken up by soybeans, which were grown in rotation with the corn. Dieldrin passed from soybeans through the agricultural food chain into the feed of domestic animals, which stored it in their fat. Meat and milk were contaminated with dieldrin throughout the United States; humans ingested dieldrin in their diet and stored it in their tissues. Dieldrin was found in the fat of almost everyone sampled in the United States. *Dieldrin caused cancer in mice and rats, even at the lowest doses tested, and the levels of dieldrin in human tissues were comparable with those in the mice and rats.*

Shell witnesses and attorneys contested each one of these facts, but their arguments were swiftly demolished by cross-examination and by contrary testimony. As in the DDT hearings, the testimony about carcinogenesis given by EDF witnesses Samuel S. Epstein and Umberto Saffiotti was especially important (Epstein, 1975).

In the end, Shell was sunk by their own data. Their research laboratory had tested dieldrin in mice at ever-lower doses, hoping to find a threshold below which there would be no effect, but they found an increased incidence of tumors even at the lowest dose tested. They had tried to find biochemical or physiological evidence that mice were in some ways unique, but they failed again. They tested dieldrin in rats, only to find that it was carcinogenic at the lowest dose tested. *As with DDT, no safe threshold for dieldrin could be demonstrated.* The hearings provided a clear demonstration of the superiority of EPA's adversarial procedures, which required competent and relevant testimony, rights of cross-examination, and full disclosure of all data relied on by either party, over the informal procedures used in other countries in which the claims of the manufacturer were accepted without critical scrutiny.

In his final recommendation, the administrative law judge accepted virtually all of EDF and EPA's arguments and recommended suspension and cancellation of all registrations of A/D. This recommendation was affirmed by Russell Train, administrator of EPA, in his final decision.

All uses of A/D in the United States were terminated by 1977. Residues of dieldrin in the human food supply then declined rapidly. Levels in human tissues and in the environment declined more slowly, with wild-life deaths from dieldrin poisoning continuing until at least 1986. Shell closed its research laboratory and stopped producing dieldrin at its last manufacturing plant in Venezuela. Aldrin and dieldrin were included in 2001 Stockholm Convention on Persistent Organic Pollutants (POPs Treaty), and they are no longer manufactured anywhere in the world. Unlike DDT, there are no cranks arguing for their reintroduction or spreading lies about the reasons for their demise.

The suspension and cancellation of A/D were among the greatest achievements of the environmental movement during the 1970s (McCray, 1977). A/D were second only to DDT/DDE in causing environmental damage in the United States (in Europe, they were first) and *posed perhaps the highest risk among all pesticides to human health. EDF played a leading role in this great achievement.* We were somewhat disappointed that the final decision was based entirely on cancer risk and that the hazards of aldrin and dieldrin to wildlife, which we had worked hard to prove, were never recognized in an administrative decision. The carcinogenicity of dieldrin has been documented in detail in the scientific literature (Epstein, 1975), but its environmental effects in the United States have never been comprehensively reviewed (Nisbet, 1988).

HEPTACHLOR/CHLORDANE: ANOTHER SUSPENSION MID-STREAM

By Ian C. T. Nisbet

EPA announced cancellation of all registrations of heptachlor/chlordane (H/C) in November 1974. These two pesticides were considered together because they had the same manufacturer (Velsicol Chemical Corporation), they were chemically similar, and they were found together in the environment. Like DDT and A/D, H/C were persistent, mobile, and retained in animal tissues, residues were found in wildlife far from the point of application, and they were ubiquitous in human tissues. The only major remaining use of chlordane by 1975 was on soil around and underneath buildings to control subterranean termites.

Although heptachlor had caused major wildlife damage in the 1950s, hazards to wildlife were not a major issue by 1975 and EDF—still busy with the mirex hearings (see below)—played a less active role than in the hearings on DDT and A/D. I served as scientific coordinator for EPA throughout the proceedings on H/C.

After lengthy preparation, the cancellation hearings were about to begin when EPA suddenly announced suspension of all uses in July 1975, again citing "new" evidence that H/C caused cancer in laboratory animals and their residues were found in the blood and fat of virtually the entire human population of the United States. This triggered another expedited hearing that was supposed to be completed in 15 days but actually lasted from August to early December. The hearings were again run professionally and fairly by Judge Perlman, and they covered much of the ground that would have been addressed in cancellation hearings, as well as the suspension issues of cancer and residues in human tissues.

Velsicol had not made Shell's mistake of following the science wherever it might lead. Many of their studies were incomplete, inadequate, or inconclusive. They succeeded in confusing the issues sufficiently that Judge Perlman in his final opinion was "cautiously unwilling" to find that H/C were carcinogenic. He was overruled by the EPA administrator, Russell Train, who suspended both pesticides on Christmas Eve, 1975. Train exempted the most important registration, however, giving Velsicol two years to prove their contention that chlordane could be used to control termites around homes without exposing the occupants. Not surprisingly, they were unable to do so, and they negotiated "voluntary" cancellation in 1977 just as EPA was about to issue an adverse finding. Chlordane labeled for this exempted use underground against termites was diverted and sold illegally for many other uses; I found it on the shelves of garden stores as late as 1979. Although it had taken more than seven years, EDF eventually achieved its goal of stopping all uses of these two pesticides.

MIREX WILL "ERADICATE" THE IMPORTED FIRE ANT

Proceedings against another "bad actor," the pesticide mirex, overlapped those against A/D and H/C and ran longer than either. With mirex, EDF played a leading role from start to finish.

The imported fire ant had spread over much of the southern United States in the 50 years since its arrival. The ant received little attention as a minor nuisance for decades, but suddenly USDA launched a barrage of horror stories about the ant in 1957. This was followed by a massive aerial program in which millions of acres were treated with dieldrin and hepta-chlor to "eradicate" the species. Large numbers of birds, fish, reptiles, am-phibians, crustaceans, and mammals, including farm animals and pets, were killed, but the fire ant survived and even expanded its range. The program was widely condemned as an ecological disaster, became highly controversial, and inspired Rachel Carson to write *Silent Spring* (Carson, 1962; Tschinkel, 2006).

USDA intended to again "eradicate" the ant, this time by a 12-year, $200 million program to spread mirex by World War II bombers on 120 million acres in nine states. Mirex is highly toxic and carcinogenic, poorly studied, cannot eradicate the ant, and would do widespread envi-ronmental damage if the program were executed. Furthermore, the pro-gram would violate a variety of antipollution laws.

Mirex killed all insects, including native species of ants that compete with fire ants, thereby actually benefiting the fire ants and facilitating their spread. The fire ants also became predators of various agricultural pests, raising the question of whether they were harmful or beneficial. The "war against fire ants" was more about money than ants—federal money from USDA to chemical companies, spray planes and equipment, local politicians, and Fourth of July picnics (Tschinkel, 2006).

EDF took this case against USDA directly to the Federal District Court in Washington, DC, on August 5, 1970, seeking a preliminary injunc-tion to stop the program. The complaint was written by EDF attorney Lee Rogers, with Denzel E. Ferguson of Mississippi State University serving as chairman of a Scientists Advisory Committee. Additional support came from scientists from Harvard, Cornell, Texas A&M, and Stony Brook universities, as well as the U.S. Department of the Interior.

USDA's motion to dismiss was denied, as was EDF's request for an injunction. USDA scaled back the program from "eradication" to "control" and launched a reduced mirex spray program. Pesticide regulation had shifted to EPA in 1970, and on April 4, 1973, EPA announced its intent to cancel mirex. The hearings began on July 11 with EDF attorneys William A. Butler and Jacqueline M. Warren representing EDF and eight other environmental organizations. Defending mirex was its maker, Allied Chemical Corporation, allied with USDA (no surprise there). The hearings droned on for two years with 100 witnesses filling 13,000 pages of transcript. EPA's decision on December 26, 1976, was to "phase out" mirex over 18 months, and mirex was banned in 1978, only eight years after EDF's original litigation. *The great war against fire ants was won by the fire ants, great profits were won by agrichemical interests, and EDF won the war against mirex.*

In nine years of federal litigation and administrative hearings, EDF had identified and succeeded in achieving a ban on six of the most dangerous chlorinated hydrocarbon pesticides—DDT, aldrin, dieldrin, heptachlor, chlordane, and mirex. All six were extremely destructive to wildlife, and all six posed a serious cancer hazard for humans. Those six were among the "dirty dozen" that were banned or tightly restricted by the POPs Treaty signed by 151 nations in Stockholm, Sweden, in 2001. EDF litigation had banned the worst of these chemicals in the United States, the "dirty half-dozen," 23 years ahead of the international POPs Treaty.

13

The DDT Wars: Four Great Victories

Nearly five decades ago a group of volunteer scientists and citizens launched a campaign to save birds from the ravages of DDT. They went to court at the local level, then through several states and finally to Washington, DC, overcoming legal barriers and challenging unexpected new issues along the way. By the 1970s, DDT and five other pesticides had been banned. Viewed from the 21st century, these actions produced significant and permanent accomplishments:

- Preventing cancer—Techniques and procedures for evaluating and regulating carcinogens, which followed the DDT precedents, have been adopted by international treaty.
- Citizen standing in court—The DDT case broke down the standing barrier, allowing citizens to go to court to protect their environment. It fostered the development of environmental law as we know it today.

- Recovery of the birds—Populations of iconic bird species, including the Bald Eagle, that had been decimated by DDT, have now
recovered their former abundance.
- Creation of the Environmental Defense Fund—EDF, spawned by
the "DDT wars," has grown into one of the nation's largest and
most influential environmental advocacy organizations.

DDT BAN LED TO GREATER CANCER PREVENTION

Top authorities in chemical carcinogenesis testified that DDT caused
cancer in laboratory animals and that it was, therefore, a possible carcinogen in humans. The precedents set by DDT for identifying and regulating
carcinogens then became the basis for banning another five dangerous
chlorinated hydrocarbon pesticides: aldrin, dieldrin, heptachlor, chlordane, and mirex (see Chapter 12). *EDF had established a very high standard for protection of public health against these carcinogens*, as confirmed
by two EPA administrators.

In 2001 the Stockholm Convention on Persistent Organic Pollutants
(the POPs Treaty) was signed by 151 nations to ban the "dirty dozen,"
which included all of the "dirty half-dozen" singled out and banned
thanks to EDF's actions 23 years earlier. There was one exception to the
total bans: DDT could be used for only malaria control. In 2009, nine additional POPs were added to the list. By 2013, 179 nations were party to
the POPs Treaty, although the United States has not yet ratified it.

As had been outlined by EDF witnesses on Long Island in 1966
and in the Wisconsin hearing of 1969, the basis for the POPs Treaty
was the same four chemical and biological properties of DDT: It is
a very stable (persistent) compound; it travels to places remote from
the site of application; it is lipid-soluble and is therefore absorbed by
living organisms; and it is a biologically active (toxic) material that
can affect those organisms. The chemicals barred by the POPs Treaty
shared these properties, making them uncontrollable once released
into the environment. *Clearly, EDF's actions in the 1960s and 1970s led*

the world in bringing these dangerous chemicals under international control.

The difficulty of identifying human carcinogens is considered in detail in the World Health Organization report of 2011. Presumably all humans worldwide are contaminated with DDT residues, as well as with dieldrin, probably the strongest carcinogen of the lot. This was essentially a worldwide exposure to toxic chemicals where there were no experimental controls.

How many human cancers and deaths resulted from these chemicals during the 30 years they were in widespread use? Were there millions of cancers, or were there none? We will never know.

This raises another key question: How many cancers were prevented by EDF's actions against this dirty half-dozen in the years after the bans? They might have prevented millions of cancers, but the answer is the same: We will never know. We can be sure, however, that the campaign to save the birds had turned into the proverbial "canary in the coal mine" that also became an effort to prevent human cancer. It is safer and cheaper to keep such chemicals out of the human environment than to allow their continued use. Banning these chemicals provided a large benefit to human health worldwide.

Dr. Samuel S. Epstein, an expert on chemical carcinogenesis, has devoted his life to preventing human cancer. He is Emeritus Professor of Environmental and Occupational Medicine at the University of Illinois School of Public Health and Chairman of the Cancer Prevention Coalition. He was an important witness in the DDT and dieldrin hearings before EPA 40 years ago. Here are some of his thoughts today:

> What began as an effort to protect birds from the ravages of DDT turned into a unique opportunity to establish policies that would prevent human cancer. Even in 1970 it was known that 70% to 90% of human cancers are environmentally induced. To prevent cancer, trace amounts of cancer-causing agents in air, water, food, and the work place must be avoided. To avoid them, carcinogens must be identified, and the hearings before the EPA on DDT and dieldrin provided an excellent forum for doing so.

Carcinogens cannot be identified using human subjects. We cannot demand such tests, but instead must depend on standardized high-dosage tests in rodents. Such tests identify carcinogens with great accuracy, and we must keep such materials out of the human environment. William Ruckelshaus of EPA fully recognized that DDT and dieldrin were carcinogens. He banned them because he was unwilling to risk human exposure to these materials. It was a great advance in cancer prevention, and he and the EDF science team that pursued the case deserve great credit. Cancer prevention saves lives, trillions of dollars in avoidable medical costs, and avoids the immense trauma of preventable cancers (Epstein, 2011).

Unfortunately, in the intervening years since these carcinogenic pesticides were identified and banned, the incidence of many non-smoking cancers has risen, the budget of the National Cancer Institute has risen substantially, yet the percent of the NCI budget devoted to cancer prevention has fallen by more than half. The American Cancer Society also has given inadequate emphasis to cancer prevention. A golden opportunity to conquer cancer is being lost (Epstein, 2011).

DDT PRECEDENTS LAUNCHED ENVIRONMENTAL LAW

It was in 1969, in Michigan, when Joseph L. Sax, the dean of environmental law, gave EDF strategic advice on how to tackle the DDT problem and overcome the legal obstacles that were then blocking citizen actions. Lack of standing would prevent EDF from challenging USDA's failure to enforce the law (FIFRA). At that time, USDA as the regulatory agency governing pesticide use would (supposedly) protect us from environmental harm. The public had no say regarding its own safety. That was the problem!

Joe Sax, now Emeritus Professor of Law at the University of California at Berkeley, offered the following observations in 2012:

The citizen-initiated DDT litigation was among the first to bring environmental citizen standing to public attention. The court cases on DDT were very important, indeed path breaking, and the DDT litigation was truly the most important effort that opened the courts to consideration of environmental issues. Although DDT was not the only early environmental issue in the courts, it was probably the most influential in that it went up against the powerful agriculture/ chemical industries. Rachel Carson's *Silent Spring* powerfully created a broad, if largely silent, constituency for the fight against chlorinated hydrocarbon pesticides, and EDF's DDT litigation carried that fight into law and policy. *Environmental law as we know it today stands on the shoulders of the citizen-standing decisions of the DC Court of Appeals in the EDF litigation against DDT.*

William H. Rodgers, a professor of law at the University of Washington School of Law in Seattle since 1967, was also a member of the EDF board of trustees in the early 1970s. He stated the following:

With the DDT cases of 1967 to 1972 EDF pioneered the use of the judicial system to give voice to expert scientific opinion in the formulation of public policy. By insisting on judicial rules of evidence, non-expert opinions of salesmen, lobbyists and others without expertise were excluded, leading to decisions based on the best available science. All too often various hearings and inquiries are based on a wide assortment of views that end up being more like a popular vote than a wise and informed decision.

The lasting legacy of the EDF campaign against DDT was the insertion of "best available science" requirements embedded in many of the laws that followed and are still with us today. These include the National Environmental Policy Act of 1969 (NEPA), which puts science, particularly the life sciences, into the mandates of all federal agencies. There are twelve "best available science" clauses in the Marine Mammal Protection Act of 1972, and eight more in the 1973 Amendments to the Endangered Species Act. Even the 1976 Magnuson Fisheries Conservation Act was bitten by the "best science" bug. Linking environmental protection to "best available science" has proven durable in practice and invaluable in policy decision-making, still with us 40 years after the EDF litigation against DDT.

REMOVE THE THREAT AND THE BIRDS WILL RECOVER

The successful EDF campaign proved beyond a reasonable doubt that DDT was causing serious damage to wildlife, especially predatory and fish-eating birds. This placed DDT in direct violation of FIFRA, and no label change could avoid that damage.

Almost all the birds that suffered declining reproductive success and diminishing population numbers for up to two decades following World War II rebounded after the 1972 DDT ban. Populations were increasing by the late 1970s, and by the year 2000, most of these species had recovered to pre-DDT population levels. The recovery adds one more layer of evidence that DDT was the major cause of the declines. With some species the recovery was stunning.

Following the DDT ban in 1972, Bald Eagle numbers in the lower 48 states increased steadily from 417 or 487 pairs (two different estimates) in 1963 to 9,789 pairs by official national count in 2006 and more than 11,000 pairs in 2007, as estimated by the Center for Biological Diversity. That's about 25 times (2,500%) as many eagles as there were before the DDT ban—a remarkable recovery (Figs. 13.1 and 13.2). In the Chesapeake Bay region, a gathering area for East Coast eagles, Bald Eagles increased nearly 11-fold, from 60 to 646 pairs, between the early 1970s and 2001 (Watts et al., 2008). In Ontario, Bald Eagle reproduction rose from 0.46 young per breeding pair in 1974 to 1.12 by 1981 (Grier, 1982). *There are*

Figure 13.1 Three newly hatched Bald Eagle chicks in the nest.
Permission from Arthur Morris, http://www.birdsasart.com.

now about 25 times as many Bald Eagles in the continental United States as there were before the DDT ban.

Bald Eagles have become accustomed to human habitation and sometimes breed within sight of it; no wonder so many people are seeing Bald Eagles these days. The Bald Eagle was listed as endangered under the

Figure 13.2 Bald Eagle nest with two chicks, photo by Mike Hamilton, permission with thanks.

Endangered Species Act (ESA) until 2007, when it was removed with fanfare and celebration. America's national bird had returned.

The species that received the most attention, research, and assistance,
the Peregrine Falcon, was also the one that suffered most from the DDT
era. Peregrines and other raptors have many friends, including ornithologists and falconers, and they sprang to the rescue. The conference in Madison in 1965 stimulated a monumental amount of research in Europe and
North America, which was summarized in another conference and a 949-
page volume 20 years later in Sacramento, California (Cade et al., 1988).

In 1970, The Peregrine Fund was founded at Cornell University, with
Tom Cade as the primary initiator, for the express purpose of raising peregrines to be released into the wild (Fig. 13.3). Learning how to
induce falcons to breed in captivity and then feeding and raising them
for release were not simple tasks. During 25 years The Peregrine Fund
and three other organizations raised and released some 7,000 peregrines
in the United States and Canada, which produced about 1,200 breed-
ers by the time the birds reached reproductive age in two to three years.

Figure 13.3 Tom Cade with Edac (Cade spelled backwards), about
1975, the tiercel (male falcon) from Alaska that fathered via artificial
insemination many of the peregrines released along the Atlantic
Coastal states, 1970s and 1980s. Permission from The Peregrine Fund.

There were numerous failures along the way. In 1984 The Peregrine Fund moved to its present location on a hilltop overlooking Boise, Idaho (www .peregrinefund.org).

The decline and the subsequent recovery of the Peregrine Falcon were truly astonishing. From an estimated 2,500 to 3,000 nesting pairs in North America pre-DDT, by 1970–1975 there were only an estimated 333 nesting pairs remaining, mostly in northern Canada and Baja California. Eastern North America had three pairs, all in Quebec; there were none in the United States east of the Rockies. But by 1999 at the time of delisting from the ESA, the number of known occupied territories had grown to 1,700, with 289 in the East, all from reintroductions. Today the estimated total is about 3,000 nesting pairs, a complete recovery. In California the total rebounded from 11 known pairs in 1971 to an estimated 300 to 350 pairs currently. With eggshells returning to normal thickness, there are now about as many peregrines in North America as there were before the DDT nightmare began (White, Cade, & Enderson, 2013; Tom Cade, personal communication). (These data refer to the American subspecies, *Falco peregrinus anatum*, and not to two Arctic subspecies in northern Alaska, Canada, and Greenland.)

It has become common for peregrines to take up residence in many cities where they had not appeared before (Fig. 13.4). The birds apparently decided that people are not so dangerous after all. Peregrines historically nest on cliffs, and many city buildings resemble cliffs. Street pigeons and starlings provide an easy and endless food supply, and living in the city means they can avoid their arch-enemy, the Great Horned Owl. It is a smart move, a win–win–win situation.

It seems especially fitting that a pair of peregrines has been nesting outside the EDF office in San Francisco, and another pair nested on a ledge outside the office window of Bill Ruckelshaus on the 37th floor of a building in downtown Seattle. Yet another pair nests in downtown Boise, within sight of The Peregrine Fund facility where so many peregrines were hatched and raised for reintroduction. The birds clearly recognize their friends! Holland, which has no cliffs and historically has had almost no nesting peregrines, now has 150 to 200 pairs, all of them nesting on

Figure 13.4 A pair of peregrines raising chicks on the MetLife Building in New York City. Photo by Chris Nadareski with his permission.

manmade structures (Cade, personal communication). Several nesting city-dwelling peregrines have been continuously monitored via webcam on Internet sites and can be viewed from around the world. In 1999, the Peregrine Falcon was removed from the ESA list, a success story without precedent in ornithological history.

The decline and the recovery of the Peregrine Falcon are told in a beautifully illustrated 394-page book, *Return of the Peregrine* (Cade & Burnham, 2003). It's a "saga of tenacity and teamwork." With recovery of the peregrine now complete, The Peregrine Fund has turned its attention to aiding other raptor species in trouble around the world, including the California Condor, Aplomado Falcon, Great Philippine Eagle, Orange-breasted Falcon, and Harpy Eagle. The Peregrine Fund has developed the world's most extensive online library on birds of prey, with 25,000 books and 30,000 reprints and scientific references, available electronically to researchers worldwide. That library was most helpful in preparing *DDT Wars*.

What has the ban on DDT got to do with all this? In the words of Tom Cade and Bill Burnham (2003):

> Let there be no doubt: the banning of DDT in 1972 was the single most important action taken to ensure the survival and recovery of the Peregrine Falcon in North America. Without it, we would not have celebrated the delisting of the American Peregrine in 1999, for it made possible everything good that happened to the Peregrine in the last decades of the 20th Century.

Since the 1972 ban on DDT, Osprey populations have been rebounding, in some cases spectacularly (Henny et al., 2010). Our litigation had stopped DDT use on Long Island in 1966, and by 1975 Ospreys were recovering (Puleston, 1975). Several nationwide U.S. surveys yielded 8,000 pairs in 1981, when the recovery had already begun, and 17,500 pairs in 2001, and the number is certainly higher now. In Connecticut and Long Island between 1938 and 1942, nests averaged 1.71 young, and this number decreased to 0.07 to 0.4 young per nest in the mid-1960s. By 1976–1977, nesting success had risen to 1.2 young per nest and DDE concentrations in unhatched eggs had declined (Spitzer et al., 1978). By 1995, there were 230 breeding pairs on Long Island alone. The birds have also benefited from more favorable public attitudes toward birds of prey, along with installation of artificial nest platforms. Some Ospreys are habituated to living with humans and will occasionally nest and raise chicks above a parking lot full of cars and people.

Along the Willamette River, Oregon, the number of occupied Osprey nests grew from 12 in 1976 to 275 in 2008 (Henny et al., 2010). In the Pacific Northwest, as in other regions, DDE concentrations correlated inversely with eggshell thickness: As DDE concentrations declined, shell thickness increased. Osprey populations in the United States are probably now as high as they have ever been (Henny et al., 2010). DDT contamination was the major cause of their decline, and the ban on DDT in 1972 was the greatest reason for their recovery.

Fortunes of the Brown Pelican also improved dramatically following the 1972 banning of DDT. By 1974 in Southern California, where reproductive success had been essentially nil in 1969 and 1970, DDE concentrations were declining, eggshell thicknesses were increasing, and far more chicks were being fledged (Anderson et al., 1975). The number of young fledged per nest was 0.004 and 0.007 in 1969 and 1970, but increased to 0.065, 0.405, 0.225, and 0.922, respectively, in 1971 to 1974. Cessation of Montrose's DDT discharges through the Los Angeles sewer system in 1970 clearly played a major role. The once-scarce Brown Pelicans have since become abundant along the entire West Coast.

By the early 1980s in the southeastern and Gulf of Mexico states, Brown Pelican populations had doubled in number from their depleted status a decade earlier. They have now fully recovered to their former abundance. The 1972 ban on DDT, especially its use on cotton in these states, is the primary reason for this recovery. The Brown Pelican was removed from the Endangered Species List in 2009.

In other species, too, improvement followed the DDT ban. The reproductive success of the Bermuda Petrel has improved and its population is increasing (David Wingate, personal communication). The reintroduced population of the California Condor also has increased, despite challenges such as lead poisoning, shooting, collisions with wires, and the ingestion of DDT-contaminated sea lion corpses (from Montrose's remaining DDT pollution). The Cooper's Hawk population throughout North America has increased substantially, approaching or exceeding its former abundance. Cooper's Hawks were once called "chicken hawk" and were mercilessly shot, but they, too, like peregrines and Red-tailed Hawks, have decided that urbanization has its merits. Seattle now has about 30 pairs nesting in wooded city parks.

The Merlin, a falcon smaller than the peregrine, has expanded its range southward since the 1972 DDT ban. Almost all species that had declined from the DDE disease in Europe showed significant population increases after the early 1970s. There is no question: *The ban on DDT proved to be an enormous benefit for predatory and fish-eating birds.*

Bob Risebrough has been investigating the effects of chemical pollutants on birds for more than half a century. Nobody knows as much about the topic. He summarized the situation recently (personal communication, 2014):

> Thousands of people and many organizations have devoted great efforts to help restore the Peregrine Falcon, Bald Eagle, Osprey and several other predatory and fish-eating birds to their former abundance. This dedicated work could only have succeeded as it did because the ban of DDT in 1972 had been achieved by the Environmental Defense Fund.

The dramatic recovery of these iconic larger birds resulting from the DDT ban stands in striking and sad contrast to the severe decline in number of the smaller tropical migratory songbirds and shorebirds during the past 60 years (Nielsen & Weidensaul, 2013). Deforestation of their tropical wintering grounds, forest fragmentation in North America, and many other factors have caused these declines. I vividly recall birding in eastern North America in the late 1940s: Forests swarmed with small songbirds and an orchestra of song. Now those same forests, otherwise intact, have only a few of those species singing solos. I would guess the decline has been 80% to 90%, but I have only my memory with no data to substantiate what has happened.

EDF TODAY: BRILLIANT, INNOVATIVE, OPTIMISTIC, AND EFFECTIVE

Average Americans Can Make a Difference

EDF was founded by average scientists and citizens without financial assets in 1967, as a child of the "DDT wars." Approaching its 50th birthday, EDF has matured into one of the largest and most influential environmental advocacy organizations in America, with an increasing presence around the world. We had no political connections and knew little about building an organization, but we had goals; we were dedicated, determined, and persistent; and we got the science right. *Every new or difficult hurdle seemed to attract a new and dedicated person who knew how to surmount it.* The EDF team was well educated and highly interdisciplinary; they worked well together and became good friends. There seemed to be no end to the outstanding and talented people who kept arriving. Scientific accuracy and integrity have always been central features of EDF. We used to joke that we were "saving the world"— now EDF is doing just that.

It is not the role of this book to trace the long and diverse history of the growth and development of EDF over the past four decades, although a few features deserve special mention. From its initial emphasis on litigation, strategies and tactics are now very different. Over the years "sue the bastards" evolved into something more like "finding the ways that work," EDF's current tagline. Often it is quicker, easier, cheaper, and more effective to find common ground within an industry than it is to be confrontational. Reducing the solid waste stream or increasing energy efficiency, for example, reduces environmental impact while increasing profit margins. EDF convinced fisheries regulatory agencies and fishermen that the "catch shares" approach will benefit the fish, reduce bycatch, increase safety and profits for fishermen, and lead to sustainability for fisheries worldwide. Economic arguments, long an EDF special strength, make such approaches plausible.

EDF has tackled many of the world's greatest environmental problems, especially where a preferable alternative can be advanced. EDF played a role in solving the "ozone hole" problem and in reversing acid rain effects. EDF economist Dr. Dan Dudek advanced the "cap-and-trade" mechanism, which became part of the Clean Air Act Amendments of 1990 and significantly reduced acid rain at minimal cost.

Climate change is the world's greatest atmospheric problem and EDF's largest program. EDF has advanced many technologies and strategies that would diminish climate change impacts. Cap-and-trade would reduce carbon dioxide emissions and climate change effects, and would do it cost-effectively. Dan Dudek has convinced China to test cap-and-trade to reduce air pollution; he found "a way that works."

EDF has an active program for restoring ecosystems damaged by human activities. Economic incentives are often an integral part of the program. Damaged habitat in one area can sometimes be exchanged for improved habitat in another. Often on private land, innovative EDF programs have improved the status of several endangered species. Restoring Louisiana's wetlands not only enhances wildlife habitat but also protects its towns and cities from storms.

EDF continues to be concerned with toxic chemicals, its earliest focus, especially from a human health perspective. Air pollution and toxic chemicals take a heavy toll on our health, especially that of children. EDF has worked with Walmart to keep toxic chemicals out of their products, with (and sometimes against) utilities to reduce dangerous emissions and enhance air quality, and with EPA and New York City to reduce dangerous diesel engine exhausts. EDF has recently won important air quality victories through litigation. The true victors are healthier citizens . . . especially the children.

Dr. George M. Woodwell, ecologist, founder of the Woods Hole Research Center, and a founding trustee of EDF, summarized his thoughts:

> The legal actions aimed at controlling DDT changed the world. Scientists with plaintive voices became articulate plaintiffs, powerfully armed with information and experience. Newspapers discovered and wrote about ecology, poisons and the public welfare. The mission of government in protecting the public from such poisons became clear and the road to the enactment in the early 70's of the revolutionary environmental laws was opened wide. Those laws are now protected and strengthened by powerful conservation-law groups who follow the early model established in the DDT wars that led to the Environmental Defense Fund more than four decades ago.

Art Cooley and I have been on the EDF board of trustees continuously for 47 years. It has been an astonishing, remarkable, breath-taking ride, and we have been most privileged to witness these awesome developments.

Art has remained a vital force within EDF for all these years. His thoughts continue to resonate:

Today, forty-seven years after its founding, EDF is a thriving organization dedicated to improving the quality of the environment. It is a well-respected national and international organization with multiple offices both in the United States and overseas. With nearly 500 employees, it has combined law, science and economics in "finding the ways that work," EDF's tagline. The initial battle over DDT taught the founders and their successors not only how to improve the environment, but how to build an organization which has become the founders' and the country's legacy.

In the 19th century, Alexis de Tocqueville toured America in part to assess whether American democracy might survive. He noted a characteristic of Americans: "As soon as several Americans have conceived a sentiment or an idea that they want to produce before the world, they seek each other out, and when found, they unite. Thenceforth they are no longer isolated individuals, but a power conspicuous from the distance whose action serve as an example; when it speaks, men listen." The founders of EDF were unaware of de Tocqueville's observations, but EDF is an organizational descendant of de Tocqueville.

Today, as I walk the shore of the Pacific Ocean near San Diego, I marvel at the results of banning DDT. Brown pelicans hang majestically on the updrafts moving for some reason known only to them to where they can feed or roost. People stand and gawk at the spectacle, and I think of how much pleasure these abundant birds, once decimated, now bring to my fellow watchers. Occasionally, a peregrine falcon invades the quiet of the shore and steals a pigeon for its fledglings. Another day an osprey sits serenely, unseen by most, on a light pole above the Cove, while a musical group readies for a concert in a summer festival. And, further inland, a pair of bald eagles nests in San Diego County for the first time since the 1930's. Yes, four iconic species returned to their original abundance bring joy and excitement to the skies of San Diego. Who could ask for a better legacy? Was the DDT battle worth fighting? Absolutely!

We have often asked each other why EDF, beginning with no obvious advantages, was such a success as an organization. We did not always agree on the ingredients for success, but here are my thoughts on the matter:

- We pursued solutions to environmental problems, which is interesting and stimulating work. We ignored the boring need to "get organized" until we were forced to do so.
- Most of us were scientists, familiar with scientific literature and well connected within the scientific community.
- Financial reward was not a motivating force.
- We discovered lawyers with a method for our science to be heard by decision makers: the courts!
- The lawyers discovered scientists, giving them something to say.
- We wrote letters to editors, to congressmen, and to legislators and we educated the kids to grow up and do it right—great things to do, but unlikely to get quick action.
- Patience is not a virtue. Persistence gets action, which attracts attention and then supporters.
- We chose large and serious national environmental problems where the best science was on our side, and where we could offer solutions.
- Early on we stressed economics, joining science and law as the third leg on the stool.
- We had illusions of grandeur. We were never a local organization that became national, but from the beginning we were national in focus, even when penniless. We figured we could turn the country around.
- We tackled controversial problems with powerful adversaries, which generates friends, enemies, and media attention. Friends are the important part.

- An adversarial process stimulates competition and the desire to win.
- Go into battle to oppose with sticks and to favor with carrots. The sticks make the carrots taste better.
- When in doubt, escalate. A good offense is the best defense.
- Get the science right, every time.
- Get the science right, **every time!**

Above all, *this is a story about people who cared*, a steadily increasing team of dedicated scientists, attorneys, economists, and other citizens, most of them volunteers without special wealth or political connections, who sought to protect the environment. It began with efforts to protect birds and their ecosystems from DDT. It expanded to curtail other environmentally dangerous chemicals and additional environmental threats. These early efforts ultimately protected human health. Legal mechanisms were developed for achieving those goals. In recent decades all of these efforts have expanded immeasurably and became worldwide to include much of humanity, yielding hope that environmental quality will be protected far into the future.

Epilogue: The Last Word by President Fred Krupp

Fred Krupp was hired as executive director of EDF in 1984 and is now president. The trustees were a bit hesitant to hire Fred at the tender age of 30, but he disarmed them by promising "to do something about my age every day." He kept his promise. He has steered the ship with excellence ever since, leading EDF to become one of the world's leading science-based environmental advocacy organizations. He should have the last word to end this book:

> I've never met a man more doggedly persistent than Dr. Charles Wurster. Without Charlie's legendary perseverance, the DDT battle in the United States might not have been won, and Environmental Defense Fund might not have been born. Thankfully, Charlie and his fellow EDF founders persisted, and their signal victory on DDT opened the way for the modern environmental movement.

Having led EDF for 30 years—half my life—I still stand in awe of the founders, who set out to save America's great birds of prey and simply would not quit, despite the forces and setbacks so vividly recounted in this book. They had to overcome enormous obstacles, and they made it happen.

Both Charlie and Art Cooley have served on the EDF Board of Trustees without interruption for 47 years and counting. They witnessed the gradual transition from a combative EDF whose informal slogan was "Sue the bastards" to an equally ambitious but more pragmatic organization whose motto today is "Finding the ways that work."

Pragmatism does not mean compromise. EDF's people today are as passionate about the environment as in 1967, and our goals are real game changers, like ushering in a new low-carbon clean energy economy worldwide. But over the years we have become more adept at harnessing market forces to protect the environment.

The seeds of that transition were planted long before my arrival in 1984. EDF began engaging Ph.D. economists in the 1970s to work alongside the scientists and attorneys, and the combination has proved invaluable in designing durable solutions.

After all, as I wrote in *The Wall Street Journal*, "Behind the waste dumps and dams and power plants and pesticides that threaten major environmental harm, there are nearly always legitimate social needs, and . . . long-term solutions lie in finding alternative ways to meet those underlying needs."

The EDF tradition of finding alternatives began at the very start. As dogged as Charlie was, he and his colleagues didn't just say "No" to DDT. They showed that there were other options available to meet the country's underlying needs.

The situation was and is different in some parts of the world. The U.S. ban never addressed the export of DDT or its use in countries plagued by malaria. Even today, EDF and others support the indoor spraying of small quantities of DDT in developing countries where malaria is spread by indoor-dwelling mosquitoes, until viable alternatives are found—a search that is still ongoing.

Appreciating the complexities of global environmental issues is increasingly important today. Challenges such as climate change and the collapse of ocean fisheries do not observe national boundaries.

These challenges are daunting, but I remain optimistic. Of course, none of us alone can solve such problems—not the United States, not China, not the World Bank, certainly not EDF. By joining forces and engaging many others, however, we must get the job done.

We won't all agree on everything. When people raise legitimate concerns, those need to be taken up and discussed, not shouted down. As my college engineering professor Charlie Walker, a tall soft-spoken Texan, once told me, people could solve a lot more problems if we would just lower our voices.

We must engage more widely, listen more carefully, find common ground and help common sense prevail. The global stakes are huge and the time horizon is short. Only by coming together with the same determination and persistence that Charlie Wurster and his colleagues brought to the DDT battle can we solve the great environmental challenges of our time.

August 2014
New York, New York

Certificate of Incorporation

ENVIRONMENTAL DEFENSE FUND,
Incorporated

Pursuant to the Membership Corporations Law of the State of New York,
WE, the undersigned, for the purpose of forming a membership
corporation, pursuant to the Membership Corporations Law of the State of New
York, do hereby certify as follows:

1.   The name of the corporation shall be the
ENVIRONMENTAL DEFENSE FUND, Incorporated.

2.   The purposes for which the Corporation is to be formed are:

(a)   To encourage and support the conservation of the natural
resources of the United States of America.

(b)   To receive and administer funds for scientific, educational
and charitable purposes and to conduct such research and disseminate the results
of such research by means of seminars, conferences, publications, and all other
means not in violation of the laws of the United States of America or the State of
New York or otherwise inconsistent with the provisions of this Certificate of
Incorporation, in furtherance of the preservation and conservation of the natural
resources of the United States of America; and to that end, to take and hold by
bequest, devise, gift, grant, purchase, lease or otherwise, absolutely or jointly with
any other person, persons, corporation or corporations, any property, real, personal
or mixed, tangible or intangible, or any undivided interest therein, without limitation
as to amount or value; to sell, convey, or otherwise dispose of any such property and
to invest, reinvest or deal with the principal or the income thereof in such manner as,
in the judgment of the Trustees, will best promote the purposes of the Corporation,
without limitation, except such limitations, if any, as may be contained in this Certificate of Incorporation, the By-Laws of the Corporation, or any laws applicable
thereto, and any limitations which might be contained in the instrument under which
such property is received by the Corporation.

(c)   The Corporation shall be conducted as an "Exempt Organization"
within the meaning of the provisions of §501(c)(3) of the Internal Revenue Code, as
amended, and the regulations promulgated thereunder, and no part of the activities
of the Corporation shall be:

(1)   attempting to influence legislation by propaganda
or otherwise; or

(2)   directly or indirectly participating in (including
the publication or distribution of statements) any
political campaign on behalf of or in opposition to
any candidate for public office; or

(3)   to have any objectives or engage in any activities
which characterize it as an action organization as
defined in Regulation 1.501(c)(3) of the Internal
Revenue Code and Regulations relating to "Exempt
Organizations" under §501(c)(3) of the Internal
Revenue Code, as amended; or

(4)   the operation of any institution named in §11 of the
Membership Corporations Law of the State of New York,
nor engage in any activities which under said section
would require the approval of any State or Local Board

Appendix 1. Certificate of Incorporation, Environmental Defense Fund,
October 6, 1967, pages 1 and 4.

DENNIS PULESTON   Meadow Lane
                                   Brookhaven, New York 11719

ARTHUR P. COOLEY   Durkee Lane
                                   East Patchogue, New York 11772

ROBERT BURNAP   21 Old Dam Road
                                   Fairfield, Connecticut 06430

H. LEWIS BATTS, Jr.   Angling Road
                                   Kalamazoo, Michigan 49001

    IN WITNESS WHEREOF, we have each made, subscribed and acknowledged this Certificate of Incorporation on the sixth day of October, in the year of Our Lord, 1967.

VICTOR JOHN YANNACONE, jr.   CAROL A. YANNACONE

CHARLES F. WURSTER, Jr.   GEORGE M. WOODWELL

ANTHONY S. TAORMINA   ROBERT E. SMOLKER

DENNIS PULESTON   ARTHUR P. COOLEY

ROBERT BURNAP   H. LEWIS BATTS, Jr.

STATE OF NEW YORK   )
) ss:
COUNTY OF SUFFOLK   )

    On this sixth day of October, 1967, before me personally came and appeared:
VICTOR JOHN YANNACONE, jr., CAROL A. YANNACONE, CHARLES F. WURSTER, GEORGE M. WOODWELL, ANTHONY S. TAORMINA, ROBERT E. SMOLKER, DENNIS PULESTON, ARTHUR P. COOLEY, ROBERT BURNAP and H. LEWIS BATTS,
each to me known to be the individuals who subscribed the foregoing Certificate of Incorporation and each severally duly acknowledged unto me that they did so subscribe it.

VICTOR J. YANNACONE
NOTARY PUBLIC, State of New York
No. 52-9763625
Qualified in Suffolk County
Commission Expires March 30, 1968

-page 4-

Appendix 1. *Continued*

IS MOTHER'S MILK FIT FOR HUMAN CONSUMPTION?

Nobody knows. But if it were on the market it could be confiscated by the Food and Drug Administration. Why? **Too much DDT.** We get it from the food we eat. It's in mother's milk, and in the body of virtually every animal on Earth — including man. DDT kills birds and fish, interferes with their reproduction, decimates their populations. It causes cancer in laboratory test animals, and people killed by cancer carry more than twice as much DDT as the rest of us. **Nobody knows for sure what DDT is doing to humans.** But who wants to wait around to find out?

That's what this country is doing. Waiting. There's been a lot of talk, but little action. You heard DDT was banned. It wasn't. Those were just empty headlines. DDT is still being used, despite acceptable alternatives.

Intolerable? Of course. It is also **illegal.** Did you know that? EDF knows it. Two big federal agencies that are supposed to protect us are not doing their job. EDF has taken them to court to see that they do.

EDF goes to court to protect the environment.

WHAT IS EDF?

EDF is the Environmental Defense Fund, Inc., a nationwide coalition of scientists, lawyers, and citizens dedicated to the protection of environmental quality through legal action and through education of the public.

WHAT IS SPECIAL ABOUT EDF?

EDF sues environmental offenders and gets action—faster than by lobby, ballot box, or protest.

EDF is a national organization, and can go to court anywhere in the country.

EDF will consider any kind of environmental case, and will tackle any offender—including the federal government.

EDF intensively prosecutes a limited number of carefully chosen cases for maximum effectiveness.

EDF works to set precedents in environmental law while solving specific environmental problems.

WHAT HAS EDF DONE?

Persistent pesticides: EDF's vigorous and effective attack on DDT in Michigan, Wisconsin, and now in Washington, D.C., has alerted this country and the world to the disastrous effects of DDT contamination of the biosphere. As a result of EDF's litigation, DDT has been curbed in Michigan, Wisconsin, and several other states. Now joined by several conservation organizations, EDF is suing two federal agencies to secure immediate action against persistent pesticides on a national level.

Air pollution: In 1968 EDF filed suit against a major air polluter in Montana. Through this case EDF hopes to establish a legal precedent recognizing the fundamental right of citizens to a wholesome environment.

Wild areas: In 1969 EDF filed suit against the U.S. Army Corps of Engineers to stop construction of the Cross-Florida Barge Canal. This canal was authorized as an escape route to protect Allied shipping from enemy submarines during World War II, but is now a useless project that will benefit a few local developers while destroying the Oklawaha River, one of the few wild rivers left in the Eastern United States.

WHAT ARE EDF'S PLANS FOR THE FUTURE?

EDF is considering litigation to achieve the following objectives:

1) to prevent the threatened sonic boom and atmospheric pollution of supersonic transports (SST's);

2) to compel a swift end to lead pollution by oil and auto companies — rather than the leisurely "phasing out" now contemplated;

3) to protect our few remaining unspoiled wetlands and estuaries;

4) to defend valuable wildlife against misguided predator control programs;

5) to stop the harvesting of certain species of whales now threatened by extinction.

MEMBERSHIP IN EDF

If you like what EDF is doing, become a member. Persuade someone else to join. EDF needs your support to increase its activities while it pursues present cases to a satisfactory conclusion. Give generously. You need your environment. Your dollars may never be more effective.

To conserve its funds for environmental action, EDF offers its members no magazines, buttons, or bumperstickers. Membership will bring you EDF's newsletter and the excitement of belonging to an action organization that is getting results in the battle for environmental quality.

EDF: ONE ANTI-POLLUTION DEVICE THAT WORKS

Appendix 2. Launching EDF's public membership, "Mother's Milk" advertisement, *New York Times*, March 29, 1970.

BIBLIOGRAPHY

Anderson, DW, JJ Hickey, RW Risebrough, DF Hughes & RE Christensen (1969). Significance of chlorinated hydrocarbon residues to breeding pelicans and cormorants. *Canadian Field Naturalist*, 82, 91–112.

Anderson, DW, JR Jehl, RW Risebrough, LA Woods, LR Deweese & WG Edgecomb (1975). Brown pelicans: improved reproduction off the Southern California coast. *Science*, 190, 806–808.

Berry, B (2014). *Banning DDT: how citizen activists in Wisconsin led the way.* Wisconsin Historical Society Press.

Best, DA, et al. (2010). Productivity, embryo and eggshell characteristics, and contaminants in Bald Eagles from the Great Lakes, USA, 1986–2000. *Environmental Toxicology and Chemistry*, 29, 1581–1592.

Broley, CL (1958). The plight of the American Bald Eagle. *Audubon Magazine*, 60, 162–163.

Brulle, RJ (2013, Dec. 21). Institutionalizing delay: foundation funding and the creation of U.S. climate change counter-movement organizations. *Climatic Change.* Available at http://drexel.edu/~/media/Files/now/pdfs/Institutionalizing%20Delay%20-%20Climatic%20Change.ashx.

Burnett, LJ, et al. (2013). Eggshell thinning and depressed hatching success of California condors reintroduced to central California. *The Condor*, 115(3), 447–491.

Burnett, R (1971). DDT residues: distribution of concentrations in *Emerita analoga* (Stimpson) along coastal California. *Science*, 174, 606–608.

Cade, TJ, & W Burnham, editors (2003). *Return of the peregrine.* Boise, ID: The Peregrine Fund.

Cade, TJ, JH Enderson, CG Thelander & CM White, editors (1988). *Peregrine Falcon populations, their management and recovery.* Boise, ID: The Peregrine Fund.

Cade, TJ, JL Lincer, CM White, DG Roseneau & LG Swartz (1971). DDE residues and eggshell changes in Alaskan falcons and hawks. *Science*, 172, 955–957.

Carpenter, DO, & R Nevin (2010). Environmental causes of violence. *Physiology & Behavior*, 99, 260–268.

Carson, R (1962). *Silent spring.* Boston: Houghton Mifflin.

Chant, D (1966). *Integrated control systems.* National Academy of Sciences—National Research Council, Publication 1402, Washington, DC.

Colburn, T, D Dumanoski & JP Myers (1996). *Our stolen future.* New York: Penguin.

Conney, AH (1967). Pharmacological implications of microsomal enzyme induction. *Pharmacological Reviews,* 19, 317–366.

Craig, PP, & E Berlin (1971). The air of poverty. *Environment,* 13, 2–9.

Curley, A, & R Kimbrough (1969). Chlorinated hydrocarbon insecticides in plasma and milk of pregnant and lactating women. *Archives of Environmental Health,* 18, 156–164.

DeBach, P (1964). *Biological control of insect pests and weeds.* New York: Reinhold Publishing Corp.

Drum, K (2013, Jan./Feb.). America's real criminal element: lead. *Mother Jones,* 28–62 (not continuous).

Dunlap, T (1981). *DDT, scientists, citizens, and public policy.* Princeton, NJ: Princeton University Press.

Edwards, JG (2004). DDT: a case study in scientific fraud. *Journal of American Physicians and Surgeons,* 9, 83–88.

Epstein, SS (1970). Control of chemical pollutants. *Nature,* 228, 816–819.

Epstein, SS (1972). Environmental pathology, a review. *American Journal of Pathology,* 66, 352–373.

Epstein, SS (1975). The carcinogenicity of dieldrin. *The Science of the Total Environment,* 4, 1–52.

Epstein, S (2011). *National Cancer Institute and American Cancer Society; criminal indifference to cancer prevention and conflicts of interest.* Xlibris.

Fry, DM, & CK Toone (1981). DDT-induced feminization of gull embryos. *Science,* 213, 922–924.

Gelbspan, R (1998). *The heat is on: the climate crisis, the cover-up, the prescription.* Cambridge, MA: Perseus Books.

Gress, F, RW Risebrough, DW Anderson, LF Kiff & JR Jehl (1973). Reproductive failures of Double-crested Cormorants in Southern California and Baja California. *The Wilson Bulletin,* 85, 197–208.

Grier, JW (1982). Ban of DDT and subsequent recovery of reproduction in Bald Eagles. *Science,* 218, 1232–1235.

Harrison, HL, et al. (1970). Systems studies of DDT transports. *Science,* 170, 503–508.

Hayes, WJ, WF Durham & C Cueto (1956). The effect of known repeated oral doses of chlorophenothane (DDT) in man. *Journal of the American Medical Association,* 162, 890–897.

Heath, RG, J Spann & JF Kreitzer (1969). Marked DDE impairment of Mallard reproduction in controlled studies. *Nature,* 224, 47–48.

Henny, CJ, RA Grove, JL Kaiser & BL Johnson (2010). North American osprey populations and contaminants: historic and contemporary perspectives. *Journal of Toxicology and Environmental Health, Part B,* 13, 579–603.

Hickey, JJ (1969). *Peregrine falcon populations: their biology and decline.* Madison: University of Wisconsin Press.

Hickey, JJ, & DW Anderson (1968). Chlorinated hydrocarbons and eggshell changes in raptorial and fish-eating birds. *Science,* 162, 271–273.

Hickey, JJ, & LB Hunt (1960). Initial songbird mortality following a Dutch elm disease control program. *Journal of Wildlife Management*, 24, 259–265.

Hickey, JJ, JA Keith & FB Coon (1966). An exploration of pesticides in a Lake Michigan ecosystem. *Journal of Applied Ecology*, 3(suppl), 141–154.

Huffaker, CB, editor (1971). *Biological control*. New York: Plenum Press.

Huffaker, CB, & AP Gutierrez (1999). *Ecological entomology*, 2nd edition. New York: John Wiley and Sons.

Innes, JRM, et al. (1969). Bioassay of pesticides and industrial chemicals for tumorigenicity in mice: a preliminary note. *Journal of the National Cancer Institute*, 42, 1101–1114.

Kiff, LF (1988). Changes in the status of the Peregrine in North America: an overview. In TJ Cade et al., editors, *Peregrine falcon populations, their management and recovery* (pp. 123–139). Boise, ID: The Peregrine Fund.

Kinkela, D (2011). *DDT and the American century*. Chapel Hill: University of North Carolina Press.

Krahn, MM, et al. (2007). Persistent organic pollutants and stable isotopes in biopsy samples (2004/2006) from Southern Resident killer whales. *Marine Pollution Bulletin*, 54, 1903–1911.

Krahn, MM, et al. (2009). Effects of age, sex and reproductive status on persistent organic pollutant concentrations in "Southern Resident" killer whales. *Marine Pollution Bulletin*, 58, 1522–1529.

Laws, E, A Curley & F Biros (1967). Men with intensive occupational exposure to DDT. *Archives of Environmental Health*, 15, 766–775.

Leopold, EB, & HW Meyer (2012). *Saved in time: the fight to establish Florissant Fossil Beds National Monument, Colorado*. University of New Mexico Press, Albuquerque.

Loucks, OL, & AR Leavitt (1999), Natural Science Foundation of Sustainability: health and integrity of resources. In OL Loucks et al., editors, *Sustainability perspectives for resources and business* (pp. 23–40). Boca Raton, FL: CRC Press.

Macek, KJ (1968a). Reproduction in brook trout fed sublethal concentrations of DDT. *Journal of the Fisheries Research Board of Canada*, 25, 1787–1796.

Macek, KJ (1968b). Growth and resistance to stress in brook trout fed sublethal levels of DDT. *Journal of the Fisheries Research Board of Canada*, 25, 2443–2451.

Macek, KJ, & S Korn (1970). Significance of the food chain in DDT accumulation by fish. *Journal of the Fisheries Research Board of Canada*, 27, 1496–1498.

McCray, LE (1977). Mouse livers, cutworms and public policy. EPA decision making for the pesticides aldrin and dieldrin. In National Research Council, *Committee on Environmental Decision Making in the EPA: Case studies* (pp. 59–105). Washington, DC: National Academies Press.

McLean, L (1967). Pesticides and the environment. *BioScience*, 17, 613–617.

Mosser, JL, NS Fisher & CF Wurster (1972). Polychlorinated biphenyls and DDT alter species composition in mixed cultures of algae. *Science*, 176, 533–535.

Needleman, H (2000). The removal of lead from gasoline. *Environmental Research*, 84, 20–35.

Nielsen, J, & S Weidensaul (2013, summer). The migratory bird decline: a status report. *Bird Conservation*, 7–34.

Nisbet, ICT (1988). The relative importance of DDE and dieldrin in the decline of Peregrine Falcon populations. In TJ Cade, JH Enderson, CG Thelander & CM White, editors, *Peregrine Falcon populations, their management and recovery* (pp. 351–375). Boise, ID: The Peregrine Fund.

Oreskes, N, & EM Conway (2010). *Merchants of doubt: how a handful of scientists obscured the truth on issues from tobacco smoke to global warming*. New York: Bloomsbury Press.

Palmer, K, S Green & MS Legator (1973). The dominant-lethal effect of p,p'-DDT in rats. *Food Cosmetics and Toxicology*, 11, 53–62.

Peakall, DB (1967). Pesticide-induced enzyme breakdown of steroids in birds. *Nature*, 216, 505–506.

Pooley, E (2010). *The climate war, true believers, power brokers, and the fight to save the Earth*. New York: Hyperion.

Porter, RD, & SN Wiemeyer (1969). Dieldrin and DDT: effects on sparrow hawk eggshells and reproduction. *Science*, 165, 199–200.

Puleston, D (1975). Return of the osprey. *Natural History*, 84, 52–59.

Radomski, JL, WB Deichmann & EE Clizer (1968). Pesticide concentrations in the liver, brain and adipose tissue of terminal hospital patients. *Food Cosmetics and Toxicology*, 6, 209–220.

Ratcliffe, DA (1967). Decrease in eggshell weight in certain birds of prey. *Nature*, 215, 208–210.

Risebrough, RW (1968). Pesticides: transatlantic movements in the Northeast trades. *Science*, 159, 1233–1236.

Risebrough, RW (1986). Pesticides and bird populations. In RF Johnston, editor, *Current ornithology* (pp. 397–427). New York: Plenum Press.

Risebrough, RW, P Rieche, SG Herman, DB Peakall & MN Kirven (1968). Polychlorinated biphenyls in the global ecosystem. *Nature*, 220, 1098–1102.

Risebrough, RW, FC Sibley & MN Kirven (1971). Reproductive failure of the Brown Pelican on Anacapa Island in 1969. *American Birds*, 25(1), 8–9.

Rogers, ML (1990). *Acorn days, the Environmental Defense Fund and how it grew*. New York: Environmental Defense Fund.

Rudd, RL (1964). *Pesticides and the living landscape*. Madison: University of Wisconsin Press.

Saffiotti, U, et al. (1970, April 22). *Evaluation of environmental carcinogens*. Report to the Surgeon General, National Cancer Institute, Bethesda, MD.

Sax, J (1970). *Defending the environment: a strategy for citizen action*. New York: Alfred A. Knopf.

Schreiber, RW, & RW Risebrough (1972). Studies of the Brown Pelican, I. Status of Brown Pelican populations in the United States. *The Wilson Bulletin*, 84, 119–135.

Short, P, & T Colburn (1999). Pesticide use in the U.S. and policy implications: a focus on herbicides. *Toxicology and Industrial Health*, 15, 240–275.

Sladen, WJL, CM Menzie & WL Reichel (1966). DDT residues in Adelie Penguins and a Crabeater Seal from Antarctica: ecological implications. *Nature*, 210, 670–673.

Smith, RF, & R van den Bosch (1967). Integrated control. In WW Killgore & RL Doutt, editors, *Pest control* (pp. 295–340). New York: Academic Press.

Spitzer, PR, RW Risebrough, W Walker, R Hernandez, A Poole, D Puleston & ICT Nisbet (1978). Productivity of Ospreys in Connecticut–Long Island increases as DDE residues decline. *Science*, 202, 333–335.

Tarjan, R, & T Kemeny (1969). Multigeneration studies in mice. *Food Cosmetics and Toxicology*, 7, 215–222.

Tomatis, L, V Turusov, N Day & RT Charles (1972). The effect of long-term exposure to DDT on CF-1 mice. *International Journal of Cancer*, 10, 489–506.

Tschinkel, WR (2006). *The fire ants*. Cambridge, MA: Harvard University Press.

van den Bosch, R (1978). *The pesticide conspiracy*. New York: Doubleday.

Wallace, GJ (1959). Insecticides and birds. *Audubon Magazine*, 61, 10–12, 35.

Wallace, GJ (1962). The seventh spring die-off of robins at East Lansing, Michigan. *Jack-Pine Warbler*, 40, 26–32.

Watts, BD, GD Therres & MA Byrd (2008). Recovery of the Chesapeake Bay Bald Eagle nesting population. *Journal of Wildlife Management*, 72(1), 152–158.

Welch, RM, W Levin & AH Conney (1969). Estrogenic action of DDT and its analogs. *Toxicology and Applied Pharmacology*, 14, 358–367.

White, CM, TJ Cade & JH Enderson (2013). *Peregrine falcons of the world*. Barcelona, Spain: Lynx Edicions.

Wiemeyer, SN, et al. (1984). Organochlorine pesticide, polychlorobiphenyl, and mercury residues in Bald Eagle eggs—1969–1979—and their relationships to shell thinning and reproduction. *Archives of Environmental Contamination and Toxicology*, 13, 529–549.

Wiemeyer, SN, & RD Porter (1970). DDE thins eggshells of captive American Kestrels. *Nature*, 227, 737–738.

Wiemeyer, SN, RD Porter, GL Hensler & JR Maestrelli (1986). *DDE, DDT & Dieldrin: Residues in American Kestrels and relations to reproduction*. Fish and Wildlife Tech. Rept. 6, US Fish & Wildlife Service, Patuxent Wildlife Research Center, Laurel, MD.

Wilkinson, PM, SA Nesbitt & JP Parnell (1994). Recent history and status of the Eastern Brown Pelican. *Wildlife Society Bulletin*, 22, 420–430.

Woodwell, GM (1967). Toxic substances and ecological cycles. *Scientific American*, 216, 24–31.

Woodwell, GM (1970). Effects of pollution on the structure and physiology of ecosystems. *Science*, 168, 429–433.

Woodwell, GM, CF Wurster & PA Isaacson (1967). DDT residues in an East Coast estuary: a case of biological concentration of a persistent insecticide. *Science*, 156, 821–824.

World Health Organization (2011). *DDT in indoor residual spraying: human health aspects*. Geneva, Switzerland: WHO Environmental Health Criteria 241.

Wurster, CF (1969a). DDT goes to trial in Madison. *BioScience*, 19(9), 809–813.

Wurster, CF (1969b). Chlorinated hydrocarbon insecticides and the world ecosystem. *Biological Conservation*, 1, 123–129.

Wurster, CF (1969c). Chlorinated hydrocarbon insecticides and avian reproduction: how are they related? In MW Miller & GG Berg, editors, *Chemical fallout: current research on persistent pesticides* (pp. 368–389). Springfield, IL: Charles C. Thomas.

Wurster, CF (1970). DDT in mother's milk. *Saturday Review*, 53, 58–59.

Wurster, CF (1971). Aldrin and dieldrin. *Environment*, 13(8), 33–45.

Wurster, CF (1972). Effects of insecticides. In N. Polunin, *The environmental future* (pp. 293–310). London: Macmillan Press.

Wurster, CF, & DB Wingate (1968). DDT residues and declining reproduction in the Bermuda Petrel. *Science*, 159, 979–981.

Wurster, CF, DH Wurster & WN Strickland (1965). Bird mortality after spraying for Dutch elm disease with DDT. *Science*, 148, 90–91.

Wurster, DH, CF Wurster & WN Strickland (1965). Bird mortality following DDT spray for Dutch elm disease. *Ecology*, 46, 488–499.

INDEX

"f." indicates material in figures and "n." indicates material in footnotes.